ENVIRONMENTAL GEOLOGY

ENVIRONMENTAL GEOLOGY

Practical Exercises

Bernard W. Pipkin

University of Southern California

David Cummings

Occidental College

PUBLISHING COMPANY

PUBLISHING COMPANY
P.O. Box 68
BELMONT, CALIFORNIA 94002
(415) 591-3505

To Richard O'Neil Stone
Teacher, scientist
and friend

Copy editor, production coordinator: Elliot Simon
Illustrations: Carol Verbeeck

Figures 1-1a, b, c, 3-1, 3-2, 3-3a, b, 4-2, 4-3, 4-4, 7-13, 8-2, 8-3a, 8-5, 8-9, 9-4, 10-3, 10-4, 11-5, and Table 3-1: Illustration by Dennis Tasa, Tasa Graphic Arts, Inc.
Typesetting: Sharon Potts & Sue Nickum
Graphics: Cheryl Baca

Printed in U.S.A.

ISBN 0-89863-058-4

Second revised printing

Introduction

Environmental geology deals with man's use of the earth. It is not only the application of science to naturally occurring problems, such as earthquakes, but also to problems that arise from man's interaction with the environment, such as mineral extraction and solid waste disposal. Environmental geology is not simply "applied geology"; it is "geology that is applicable." Applicable means collecting the proper data, analyzing it, and presenting it in a way that can be used by non-specialists and decision makers, such as planners and politicians.

Much of this manual is concerned with natural geologic processes and geologic hazards. These processes become geologic hazards when they interfere with the works of man. We believe that every topic in geology can be related to man and his engineering works. These chapters were written with this belief in mind, and they emphasize the practical and applied.

Environmental geology is useful beyond the classroom. We hope that some of the lessons learned here can be helpful to the student and teacher in the selection of homesites or in making investments. But most of all, we wish to provide students with an understanding of their geological environment that will enable them to make intelligent decisions on the complex issues that face their generation and generations to come.

Bernard W. Pipkin
David Cummings

Acknowledgments

The authors wish to thank their colleagues and students for many helpful suggestions and additions to the manual. We are particularly indebted to Dr. David L. Schleicher, U.S. Geological Survey, who wrote the chapter on environmental impact statements, to Dr. Douglas Hammond, who helped develop the exercise on water resources, and to Linda Thurn, Cindy Wiersma, and Jim Hollywood, graduate assistants at the University of Southern California, for student-testing exercises and making needed revisions. We also wish to thank our colleagues in the Geophysical Laboratories at U.S.C. for source materials and information on earthquakes.

The California Division of Mines and Geology was most cooperative in providing illustrations and granting permission to use some of their published material. We would also like to thank all those students who persevered during the very rough trials in our respective Environmental Geology laboratories. For any errors of commission we take full responsibility.

Introduction

Contents

Chapter 1

Earth Materials and Processes

Every aspect of environmental geology in one way or another involves geologic materials—rocks, minerals, and soils—and their use by man, such as for building materials, structural foundations, metals, and fossil fuels. Whereas in geologic research we are interested in the age of geologic materials, their origin and paleoecological or geochemical significance, in environmental geology we are concerned with geological materials as they are useful to man or as they pose a potential geologic hazard.

TECTONIC CYCLE

The Earth is a constantly changing dynamic system. In order to recognize this we must consider the great size of the Earth, the relatively thin crust covering the planet, which we call the continents and ocean basins, and the enormity of geologic time. Within our human time frame, the Earth may not appear especially dynamic, but over the span of thousands or millions or billions of years, the Earth is never at rest. Many cycles make up the dynamic Earth, all acting interdependently. The tectonic cycle is just one of these many cycles.

Most earth scientists today accept the **plate tectonic theory**, which states that the Earth is composed of several large plates and many smaller ones. These plates are in constant motion and are colliding, separating, or slipping past one another. The plate tectonic theory gives us a hypothesis for the formation of great mountain chains such as the Alps and Himalayas; it helps explain the distribution of volcanic activity on Earth and the distribution of strong and weak earthquakes. Many of these indicators of crustal unrest are concentrated at **plate boundaries**, of which there are three types:

Divergent (Constructive) Boundaries. These are areas along mid-ocean ridges and rises where plates are spreading and the gap created is filled with molten material from the mantle below. New oceanic crust (basalt) is formed, thus the term "constructive boundary." Examples are the Mid-Atlantic Ridge and the East Pacific Rise. They are characterized by non-explosive volcanic activity, high heat flow, and weak, shallow earthquakes (Fig. 1-1a).

Convergent (Destructive) Boundaries. Because new material is added at divergent boundaries, the Earth would get larger unless an equal amount were lost somewhere else.

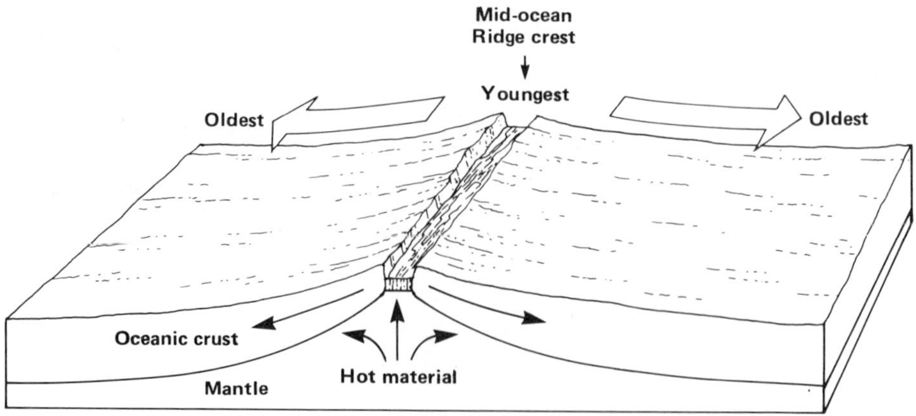

FIGURE 1-1a. Divergent plate boundary along a mid-ocean ridge. This is also known as a spreading center because new sea floor spreads out from the ridge crest.

This takes place at convergent boundaries where one rigid plate collides with and sinks beneath another to be remelted. These are areas where the Earth's crust bends downward and are the locations of great oceanic trenches and deeps. Examples are the Chile-Peru Trench, the Marianas Trench, and the Acapulco Trench. Convergent boundaries are characterized by strong earthquakes and explosive volcanoes (Fig. 1-1b).

Conservative Boundaries. These are areas where two plates are slipping past one another and the crust is neither created nor destroyed. An example is the great San Andreas fault in California where the Pacific Plate is moving northward relative to the North American Plate. These areas have moderate to strong earthquake activity (Fig. 1-1c).

Thus we see that the Earth's crust is constantly being recycled. That is, new material is being formed at the mid-ocean ridges to move outward and ultimately sink at the trenches to be remelted. We have used the term tectonic cycle to describe this recycling of the Earth's crust. It has happened many times over the 4.5-billion-year history of this planet. Scars of former trenches and conservative boundaries are to be found well in the interior of the continents.

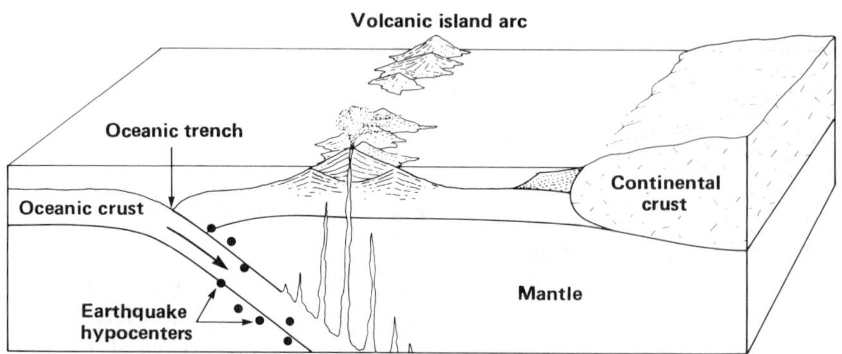

FIGURE 1-1b. Convergent plate boundary at a deep-sea trench. This is also known as a subduction zone because sea floor is consumed or "subducted" at the boundary.

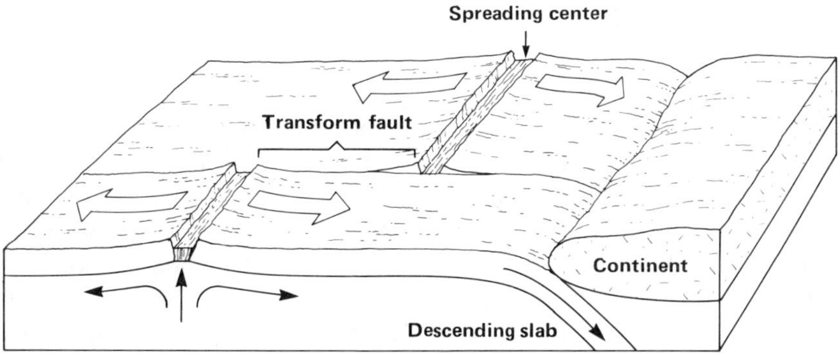

FIGURE 1-1c. Conservative plate boundary or transform fault. New crust is neither created nor destroyed along this type of boundary. The San Andreas fault in California is an example.

THE ROCK CYCLE

Just as we can describe a tectonic cycle, we can also describe a rock cycle whereby rocks form and are destroyed. As we have already seen, new rocks form at some tectonic plate boundaries from molten material. Such "igneous" rocks are broken down by weathering and erosion to form "sedimentary" rocks, composed of particles. These rocks in turn may be eroded or altered to other kinds of rocks (metamorphic) by heat and pressure. Figure 1-2 shows the cycle and history of rocks on Earth.

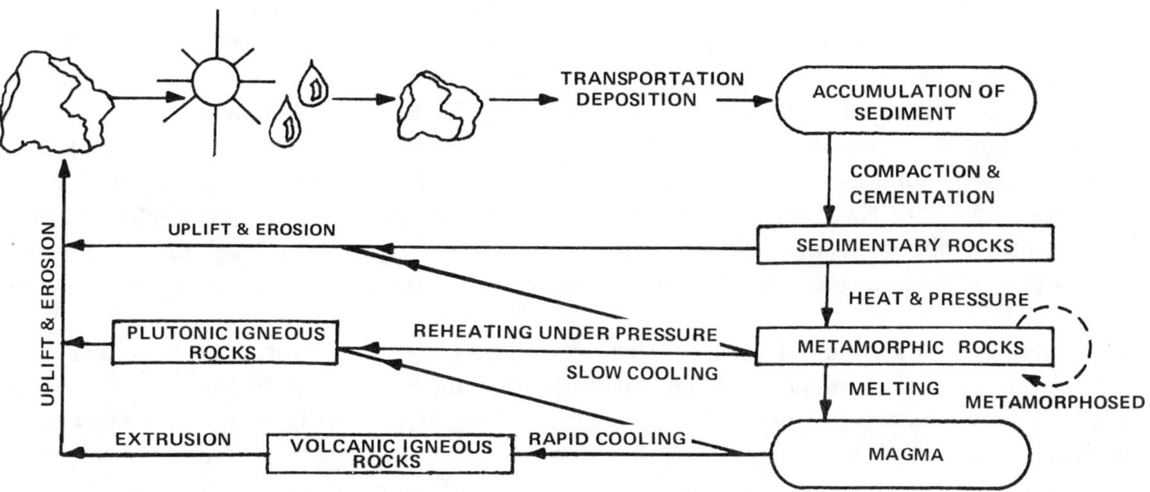

FIGURE 1-2. The rock cycle.

CLASSIFICATION OF ROCKS

Rocks are aggregates of minerals that form a significant part of the Earth's crust. Minerals are homogeneous, naturally occurring inorganic substances with a definite set of physical properties and a chemical composition that varies within set limits. As examples, there are quartz (SiO_2), the most abundant mineral, and orthoclase ($KAlSi_3O_8$), a very common mineral. These minerals may occur in pure form in nature but when they crystallize together in set proportions they form the rock granite. Similarly, when plagioclase ($CaAl_2Si_2O_8$) and pyroxene ($CaMgSiO_4$) crystallize from a melt in certain proportions, basalt results.

There are three principal classes of rocks based upon their origin: **igneous, sedimentary,** and **metamorphic**. In environmental geology, our interest is in rocks of environmental importance and we will restrict our discussions to those rocks that are most commonly involved in geologic processes hazardous to man, such as landslides, dam failures, and the like.

Rocks can also be classified on the basis of their texture and composition. Texture is defined as the size, shape, and relationship of the grains in the rock. Fine-grained rocks have very small grains not easily seen with the naked eye, and coarse-grained rocks are those in which the individual grains are easily seen without the aid of a microscope or hand lens. **Color** is a clue to the composition of some rocks, but not all.

Class I — Igneous Rocks. Igneous rocks are formed by crystallization from molten material. They consist of interlocking crystals and may be identified by their mineral content and texture. Texture reflects the mode of formation and cooling history of a particular igneous rock. Rocks that crystallize slowly from molten material (known as **magma**) within the Earth are generally coarse grained. On the other hand, rocks that form from molten material at the surface of the Earth cool rapidly, and their crystals are very small. Coarse-grained igneous rocks are known as **phanerites** (phanertic texture) and include such familiar examples as **granite** and **gabbro** (Table 1-1). Fine grained igneous rocks are known as **aphanites** (aphanitic texture) and include the common rocks **rhyolite** and **basalt**. There are rock types intermediate in texture, such as **diorite** and **andesite**. The rocks shown in Table 1-1 represent the most common types of igneous rocks. Thus the term "granitic" rock includes a host of rock types that constitute the bulk of the coarse-grained light-colored igneous rocks of the Earth's crust (Fig. 1-3). Glassy rocks such as **obsidian** (black glass) and **pumice** (spun or frothy glassy) are special types. Although they are relatively common, they present special engineering problems and have rather specialized industrial uses. **Pyroclastic** literally means "fire broken" and refers to igneous rocks formed from ash, cinders, or rock fragments blown from a volcano. Large areas of the United States are underlain by tuffs and tuffaceous rocks indicating volcanic eruptions have taken place in the past, similar to the May 18, 1980, eruption of Mount St. Helens in Washington. In fact, the evidence from thick, widespread deposits of tuff in Oregon and Washington indicates that many of these eruptions were larger and more violent than the famous eruption of 1980.

Class II — Sedimentary Rocks. Sediment is particulate matter formed by chemical and physical breakdown of material of the Earth's crust. This material is transported by a geologic agent, such as wind, running water, or glaciers, and deposited layer by layer at another site. Sedimentary rock is sediment that has become coherent or firm by pressure and cementation. Pressure is a function of depth of burial; and natural cements, such as calcium carbonate ($CaCO_3$), silica (SiO_2), and iron oxide (Fe_2O_3), cause mineral grains to adhere and make

TABLE 1-1
IGNEOUS ROCKS

	Composition		
Texture	Acidic, High SiO_2, Light-colored quartz and orthoclase*	Intermediate	Basic, Low SiO_2, Dark-colored pyroxene and plagioclase*
Coarse grained (phanerites)	Granite	Diorite Quartz diorite	Gabbro
Fine grained (aphanites)	Rhyolite (felsite)	Andesite	Basalt
Glassy	Obsidian Pumice		
Pyroclastic	Tuff	Tuff	Tuff

*Orthoclase and plagioclase are members of a larger family of minerals known as the Feldspar Group.

FIGURE 1-3. Igneous rocks (scale is 5 centimeters long). 3a. Granite showing salt-and-pepper appearance and uniform coarse texture. 3b. Diorite composed of a higher percentage of dark iron-rich minerals. 3c. Volcanic glass or obsidian exhibiting a glassy or vitreous texture. 3d. Pumice, showing the characteristic porous appearance. 3e. Tuff exhibiting pyroclastic texture. Note rock fragments incorporated in the sample. 3f. Rhyolite

the rock hard. **Sandstone**, for instance, forms by cementation of resistant minerals that are produced by erosion of granular rocks such as granite or pre-existing sandstones. The principal mineral in the average sandstone is quartz, the hardest and most resistant of the common minerals. The process whereby loose sediment becomes a sedimentary rock is shown diagramatically below:

Loose Sediment	→	**Lithification**	→	**Sedimentary Rock**
Mud, sand, gravel, etc.		Pressure from depth of burial or cementation by natural cementing agents		Shale, sandstone, conglomerate

Some sedimentary rocks form directly or indirectly from the activity of organisms. Thus many limestones are composed of skeletal material of microscopic plants or broken shells of larger marine organisms, and coal is the carbonaceous residue of great accumulations of land-plant debris. Probably the most notable feature of all sedimentary rocks is their bedding or **stratification** (Fig. 1-4).

Sedimentary rocks are divided into **clastic** and **non-clastic** subclasses. Clastic rocks are those composed of mineral or rock fragments, whereas non-clastic rocks are formed by the activity of organisms or as chemical precipitates or both. Table 1-2 outlines the major groups of sedimentary rocks. Note that clastic rocks are classified by grain size, which is a non-genetic classification and is therefore a very simple and utilitarian one (Fig. 1-5).

FIGURE 1-4. Stratification in folded sedimentary rocks, Mojave Desert. Note upfold on the left (anticline) and downfold on the right (syncline). (Photo courtesy R.O. Stone.)

TABLE 1-2
SEDIMENTARY ROCKS

Texture (grain size)	Composition	Rock Type	General Features
CLASTIC			
Coarse (over 2,000 μ)***	Gravel	Conglomerate	Rounded rock fragments in a matrix of finer material.
Medium (2,000–64 μ)	Sand	Sandstone Arkose* Greywacke**	Mainly quartz, some feldspar and minor rock fragments and other mineral grains.
Fine (64–4 μ)	Silt	Siltstone	Quartz and finely divided minerals.
Very fine (less than 4 μ)	Clay	Shale (if laminated) Mudstone (if massive)	Mostly clay minerals.
NON-CLASTIC			
Coarse to fine, crystalline	Calcite ($CaCO_3$)	Limestone	White to dark grey, effervesces in HCl.
Coarse to fine, crystalline	Dolomite ($CaMg(CO_3)_2$)	Dolomite	Grey to black, difficult to dissolve in acid.
Coarse to fine, crystalline	Halite (NaCl)	Rock Salt	Salty taste, dissolves in water.
Coarse to fine, crystalline	Gypsum ($CaSO_4 \cdot 2H_2O$)	Gypsum	Weakly soluble in water, softer than fingernail.

*More than 25% feldspar grains. **High % of rock fragments and clay minerals.
***See appendix B —— μ = micron = 0.001 millimeter

a.

b.

c.

Figure 1-5. Sedimentary rocks (scale is 5 cm long). 5a. Conglomerate with well-rounded pebbles and a sandy matrix. 5b. Silty shale showing laminations. 5c. Limestone composed entirely of shell materials. This kind of limestone is called a "coquina." 5d. Light-colored fine-grained sandstone. 5e. Arkose

d.

e.

TABLE 1-3
STRATIFICATION AND SPLITTING CHARACTERISTICS

Stratification (Bedding)	Splitting Characteristics	Thickness*
Thick	Blocky or Massive	Greater than 2 ft
Thin	Slabby	2 in.–2 ft
Very thin	Flaggy	½ in.–2 ft
Laminated	Shaly (shale)	2 mm–1 cm
Laminated	Platy (limestone and sandstone)	2 mm–1 cm
Thinly laminated	Fissile (paper thin)	Less than 2 mm

*Approximate—Note change of units from English to Metric

Bedding or stratification characteristics are extremely important because thinly bedded or layered rocks have more "defects" and are thus more likely to fail or slide than thick-bedded or massive ones. This will be discussed in more detail in Chapter 8 on Mass Wasting and Landslides.

a.

b.

c.

d.

e.

f.

FIGURE 1-6. Metamorphic rocks (scale is 5 cm long). 6a. Gneiss showing coarse, poorly developed foliation. 6b. Slate exhibiting slaty cleavage or break. 6c. Marble exhibiting uniform crystallinity and streaks of dark-colored minerals. 6d. Mica schist showing characteristic crinkly foliation planes. 6e. Phyllite showing mineral crystals on foliation surface. 6f. Quartzite

A descriptive classification of stratification and how layered rocks break is presented in Table 1-3.

Class III — Metamorphic Rocks. These are pre-existing rocks that have been changed in the solid state by heat, pressure, and chemically active fluids. The process of metamorphism results in new structures, producing a layered appearance, new minerals, and recrystallization of old minerals into larger grains. The net result of metamorphism is for a rock to have greater crystallinity (more uniform and larger crystals) and greater hardness, for example, marble that forms from limestone and slate that forms from shale. The new structure that is gained by some rocks during metamorphism is known as **foliation** and is the basis of the classification. Foliated rocks have a layered appearance and the coarseness of the layering determines the type. Thus, finely foliated rocks are knows as **slates**, moderately foliated rocks are called **schists** (schistosity consists of layered platey minerals like mica), and coarsely layered rocks are called **gneiss** (Fig. 1-6).

Non-foliated metamorphic rocks are recrystallized pre-existing sedimentary or igneous rocks. Thus limestone becomes **marble** when it is subjected to the appropriate heat and pressure, and sandstones become **quartzite**, which is virtually a solid mass of recrystallized quartz grains. Metamorphic rocks are extremely durable but the same weaknesses exist along foliation planes in metamorphic rocks as are found along bedding planes in sedimentary rocks. Failure of hill slopes and road cuts tend to take place along these planes, particularly when clayey or micaceous minerals occur on these surfaces. Table 1-4 shows the classification scheme of the predominant metamorphic rock types.

ENVIRONMENTAL IMPORTANCE

The optimal application of geologic information to environmental problems requires knowledge of earth materials, their geologic significance and their strength or engineering properties. While there are over 2,000 minerals and hundreds of rock types, just a few make up the bulk of exposed materials and are significant to man's relationship with his geologic environment. For instance, certain rocks are great underground water reservoirs and cannot be allowed to be contaminated by wastes. Other rocks are inherently weak and if fashioned into high, steep slopes they will require a great deal of maintenance. Some soil types are even known to be related to endemic diseases.

Table 1-5 lists the most common rock types and their general engineering properties. Keep in mind that these are just general guidelines and that such factors as the presence or absence of fractures, interstitial water, or deep weathering may drastically alter the condition of the rock and make it less desirable from an engineering point of view. For example, many ancient shales are durable and extremely hard and make good foundations or roadcuts, while young fractured shales are usually quite susceptible to sliding, or form soils that readily erode.

TABLE 1-4
METAMORPHIC ROCKS

Subclass		Derived from	Rock Type	Characteristics
Foliates	↑ increasing metamorphism	Granite, diorite, gabbro, coarse-grained rocks	Gneiss	Alternating light and dark layers; quartz, feldspar, mica, and garnet are common minerals.
		Most any fine-grained igneous or sedimentary rock, such as shale, basalt, rhyolite, and siltstone	Schist	Usually crenulated with shiny, parallel-oriented mica flakes; quartz and feldspar are common.
		Fine-grained rocks, shale, or mudstone	Phyllite	Parallel bands with shiny mica surfaces.
		Fine-grained rocks, shale, or tuff	Slate	From old German "to split," cleaves along closely spaced parallel planes, usually dark colored.
Non-Foliates (or faintly foliated)		Sandstone	Quartzite	Hardest of common rocks, sugary texture, glassy appearance on freshly broken surfaces.
		Limestone and dolomite	Marble	Well crystallized, may show streaks of dark-colored or oxidized minerals.
		Fine-grained rocks	Hornfels	Dark colored, variable mineralogy, usually quite hard and dense.

TABLE 1-5
ROCK TYPES FOR FOUNDATIONS AND EXCAVATIONS

Earthquake Response	Engineering Property	Rock Type
Lower	Excellent foundations, may be difficult to excavate where fresh and unfractured.	Igneous rocks Gneiss Quartzite
↑	Very good foundations, may be difficult to excavate where unfractured and fresh.	Limestone Dolomite Marble (possible sink holes and cavernous rock)
Variable	Moderately good foundations, may or may not be difficult to excavate depending upon degree of weathering and cementation.	Schists Sandstone Interbedded sandstone and shale
↓	Variable foundation conditions, easy to difficult excavation.	Clay shale Siliceous shale—generally very hard Mudstone Siltstone
Higher	Variable foundation conditions, usually easy to excavate.	Flood-plain deposits, alluvium, unconsolidated and weathered rock materials

EXERCISE I

EARTH MATERIALS AND PROCESSES

Name ______________________________

1. Draw a model of the rock cycle. Use information from the lecture, lab, and your textbook. At each point in the cycle give several examples of each rock class and the various processes that act upon them (heat, pressure, weathering, cementation).

2. Explain the distribution of active volcanoes and earthquakes on the Earth.

____

____

____

____

3. Define, sketch, or otherwise identify:
   (a) three classes of rocks

   (b) weathering and erosion

   (c) three types of plate boundaries

4. Describe briefly what is meant when we say the Earth is a dynamic system.

5. On the chart below name the rock samples in the sets given to you. It is usually best to work together in pairs and discuss results and conclusions with your lab partner. In the last column indicate any obvious economic use for the rock material, such as building stone, abrasives (pumice), or roofing materials. Finally, if the rock has any significant environmental or engineering property worthy of note, cite this feature in the last column. For instance, if the shale sample in your collection is rather soft and laminated it probably will not be capable of supporting steep slopes.

	No.	Rock Name	Color	Texture	Environmental Significance
Igneous					
Sedimentary					
Metamorphic					

Chapter 2
Topographic Maps

For environmental geologic information to be useful to the reader, it must be conveyed in an accurate and convenient manner; maps serve this purpose. Maps supply information on the surface or subsurface conditions. Maps are one of the primary tools the geologist needs to display certain basic information (topography of the area; geologic or soil conditions); they are used to plot, analyze, and interpret the data.

Many different types of maps can be made, for either general or specific purposes. The following discussion provides the fundamental background for understanding and using topographic and geologic maps. Special purpose maps will be introduced elsewhere in the laboratory manual, where direct application can be more readily seen.

TOPOGRAPHIC MAP FEATURES

A topographic map is a two-dimensional representation of the Earth's surface and illustrates both natural and man-made features. In addition to showing the configuration of the Earth's surface, maps can be used to accurately determine a location and to show both horizontal and vertical distances corresponding to the actual distances in the field. A set of standard map symbols has been developed to depict natural and man-made features (Table 2-1). Other information, such as scale and name, is shown in the margin of the map. The most common information is shown in Figure 2-1. Diagrams showing the relation of the land surface to the corresponding topographic map are shown in Figure 2-2.

Control. The accuracy (control) of both horizontal and vertical distances shown on a map depends on the accuracy of field measurements and the number of data points (stations). Most commercially prepared maps include information on the degree of accuracy as well as the number of stations. The most accurately measured stations are called **bench marks**. The elevations and distances are measured from a base station and are periodically remeasured to determine if any changes in elevation or horizontal distances occurred since the last measurements. At a lower degree of accuracy is the **level line survey**. Other lower-accuracy control stations are indicated by the type of survey.

Direction. Maps commonly are oriented so that the arrow used to designate **geographic**

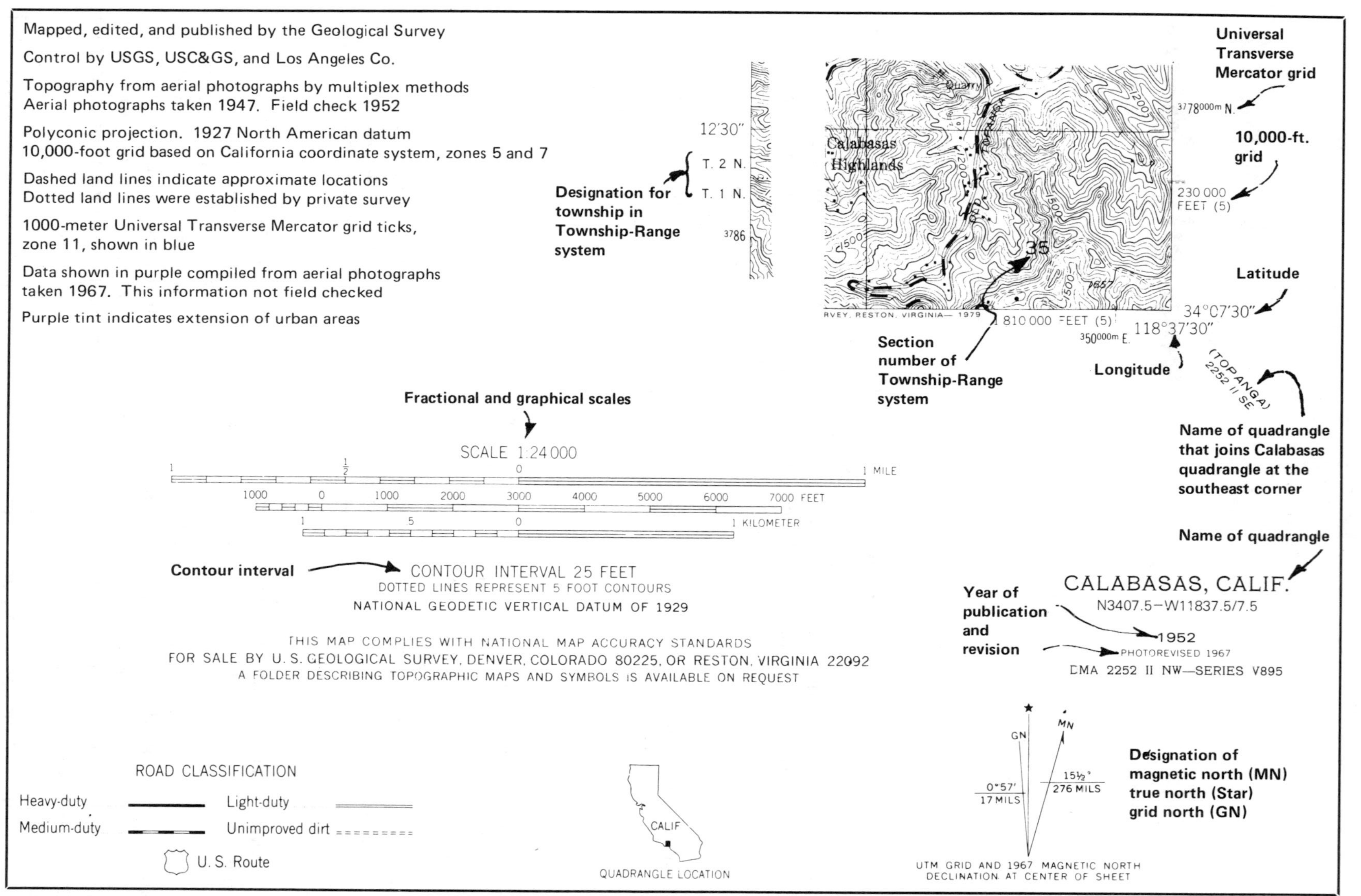

FIGURE 2-1. Information that appears in the margins of U.S. Geological Survey quadrangle maps.

TABLE 2-1
TOPOGRAPHIC MAP SYMBOLS*

Primary highway, hard surface	Boundaries: National
Secondary highway, hard surface	State
Light-duty road, hard or improved surface	County, parish, municipio
Unimproved road	Civil township, precinct, town, barrio
Road under construction, alinement known	Incorporated city, village, town, hamlet
Proposed road	Reservation, National or State
Dual highway, dividing strip 25 feet or less	Small park, cemetery, airport, etc.
Dual highway, dividing strip exceeding 25 feet	Land grant
Trail	Township or range line, United States land survey
	Township or range line, approximate location
Railroad: single track and multiple track	Section line, United States land survey
Railroads in juxtaposition	Section line, approximate location
Narrow gage: single track and multiple track	Township line, not United States land survey
Railroad in street and carline	Section line, not United States land survey
Bridge: road and railroad	Found corner: section and closing
Drawbridge: road and railroad	Boundary monument: land grant and other
Footbridge	Fence or field line
Tunnel: road and railroad	
Overpass and underpass	Index contour / Intermediate contour
Small masonry or concrete dam	Supplementary contour / Depression contours
Dam with lock	Fill / Cut
Dam with road	Levee / Levee with road
Canal with lock	Mine dump / Wash
	Tailings / Tailings pond
Buildings (dwelling, place of employment, etc.)	Shifting sand or dunes / Intricate surface
School, church, and cemetery (Cem)	Sand area / Gravel beach
Buildings (barn, warehouse, etc.)	
Power transmission line with located metal tower	Perennial streams / Intermittent streams
Telephone line, pipeline, etc. (labeled as to type)	Elevated aqueduct / Aqueduct tunnel
Wells other than water (labeled as to type) — Oil, Gas	Water well and spring / Glacier
Tanks: oil, water, etc. (labeled only if water) — Water	Small rapids / Small falls
Located or landmark object; windmill	Large rapids / Large falls
Open pit, mine, or quarry; prospect	Intermittent lake / Dry lake bed
Shaft and tunnel entrance	Foreshore flat / Rock or coral reef
	Sounding, depth curve (10) / Piling or dolphin
Horizontal and vertical control station:	Exposed wreck / Sunken wreck
Tablet, spirit level elevation — BM Δ5653	Rock, bare or awash; dangerous to navigation
Other recoverable mark, spirit level elevation — Δ5455	
Horizontal control station: tablet, vertical angle elevation — VABM Δ9519	Marsh (swamp) / Submerged marsh
Any recoverable mark, vertical angle or checked elevation — Δ3775	Wooded marsh / Mangrove
Vertical control station: tablet, spirit level elevation — BM × 957	Woods or brushwood / Orchard
Other recoverable mark, spirit level elevation — × 954	Vineyard / Scrub
Spot elevation — × 7369 × 7369	Land subject to controlled inundation / Urban area
Water elevation — 670 670	

*May vary on older maps. Courtesy of U.S. Geol. Survey.

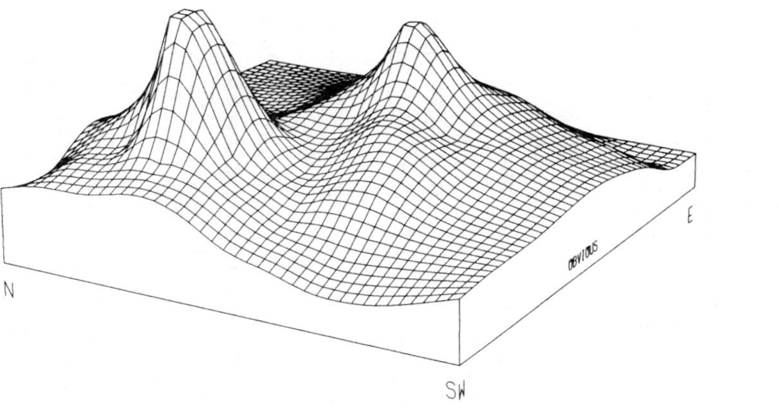

FIGURE 2-2a. Computer graphics of hilly terrain.

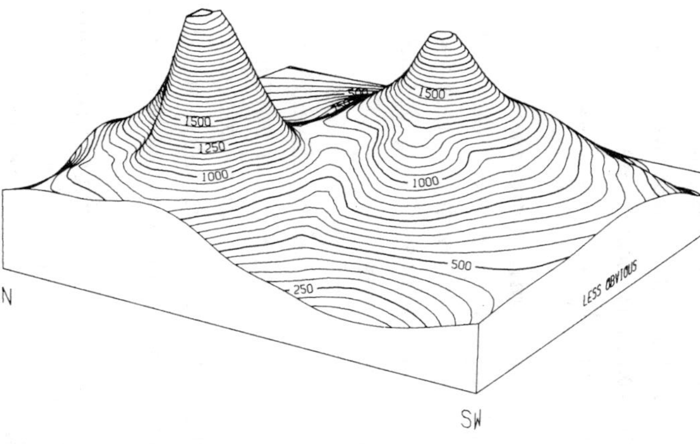

b. Same terrain with contour lines at 50-ft intervals.

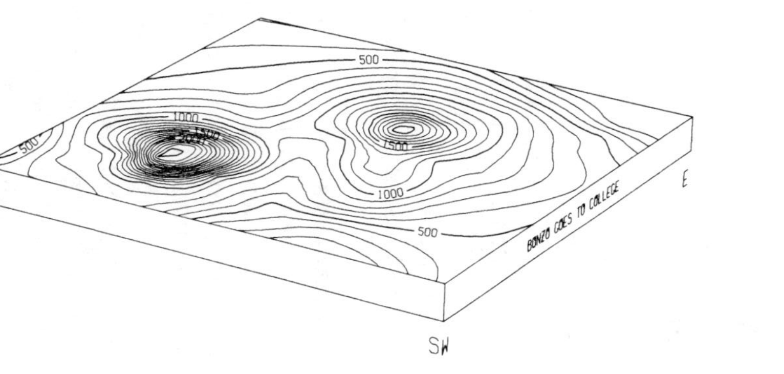

c. Oblique view of same terrain, flattened, showing contour lines but no relief.

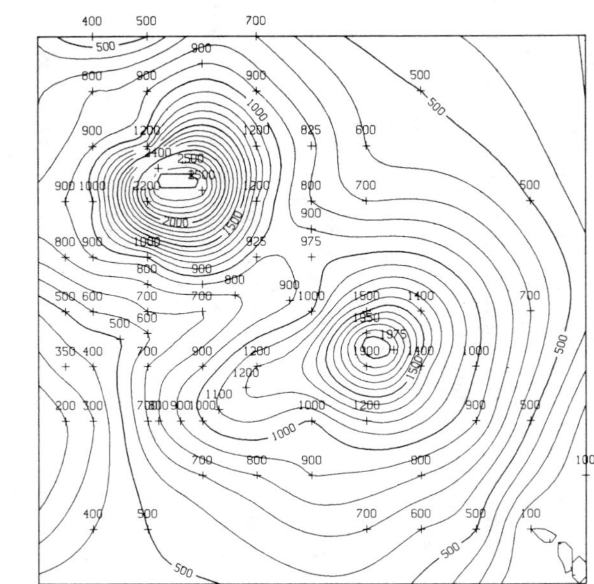

d. Contour map of same terrain, showing data points used to construct contours. (Courtesy of Jon Gaynor and Union Oil Company of California.)

north points toward the top of the map. Typically, the geographic north pole is not coincident with the **magnetic north pole** (usually determined in the field). The angular difference between geographic north (or true north) and magnetic north is called the **magnetic declination** (Fig. 2-1), which may be east or west of true north, depending on the location. The line where geographic north and magnetic north coincide is called the **agonic line**.

Direction may be described by the amount of deviation from true north. Three methods are generally used: (1) points of the compass (e.g., due north, northeast, southwest, due east); (2) azimuthal direction, where the compass directions are divided into 360°, north = 0°, east = 90°, south = 180°, west = 270°; (3) a combination of (1) and (2) whereby we might say "north 35° east," or "north 82° west"—abbreviated N35E and N82W, respectively.

Location. Most topographic maps, or **quandrangles**, are bounded by latitude and longitude lines. The boundaries of the map may be in full degrees (e.g., 1° × 2°) or in minutes (e.g. 7½′ × 7½′). There are 60 minutes of arc in one degree. The latitudes and longitudes of the quadrangles are shown at the four corners of the map and at evenly spaced intervals along the margins (Fig. 2-1). We may locate a feature by using the intersection of the latitude and longitude lines.

Most of the United States west of the Appalachians is divided into a system of **townships** and **ranges**. The townships are bounded by a **principal meridian** (longitude) and are read east or west of that meridian, and the ranges are bounded by a **base line** (latitude) and are read north or south of that line (Fig. 2-3). Each township and range is a square that is 6 miles on a side or 36 square miles. Each of the 36 one-mile squares is called a **section**. The sections are numbered consecutively as shown in Figure 2-3. Each section can be further divided into quarters and each quarter can be further divided into additional quarters. Locations are given starting with the smallest quarter, through the larger quarter, through the section number, to the township and range numbers (Fig. 2-3).

Other means of location depend on a grid system. The distances along the grid lines are measured north and south as well as east and west from some base point. On the map, the grid lines are indicated by a short dash (tick mark) at the margin of the map (Fig. 2-1). A Universal Transverse Mercator grid using 1,000-meter (1-km) intervals is a common system (Fig. 2-1). For location purposes we use the intersection of the grid lines. An arbitrary grid system may be devised for special purposes, such as engineering projects or road alignments.

Scale — Horizontal Distances. The reduction of the size of the real world to the size of a map demands that the horizontal and vertical relations of features in the field be accurately scaled down in order for the map to be a true representation of the field conditions. Thus, horizontal distances between features on the map have the same ratio as distances in the field. Maps are made so that when we measure the horizontal distance between features on them, the distance can automatically be converted to the distance between the same features in the field. For this purpose we use a **graphic scale**, which appears on the map and is ordinarily represented by a line divided into equal parts (Fig. 2-1). Each division on the scale is equivalent to a certain distance in the field. One advantage of the graphic scale is that it can be used to directly convert distance on the map to distance in the field. Another advantage is that if the size of the original map is reduced or enlarged, the graphic scale changes proportionately and the proper correspondence of distance is maintained. Only the conversion factor is changed.

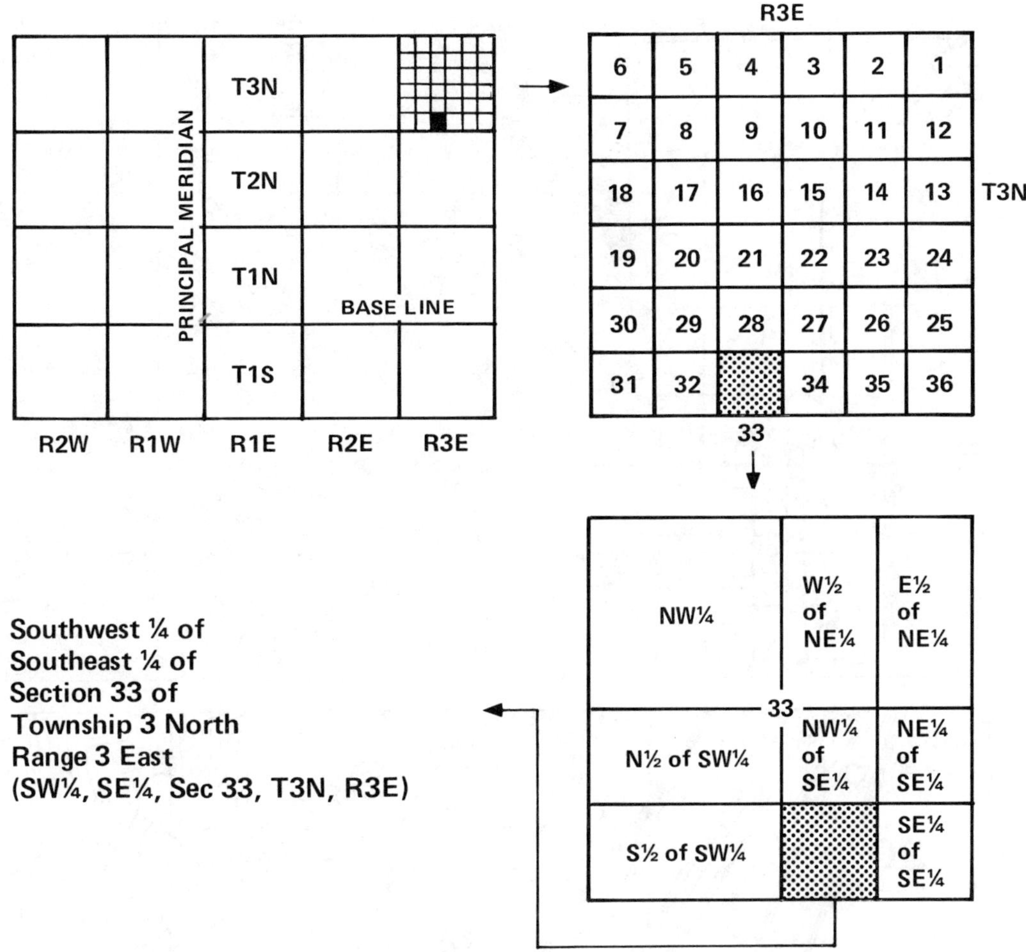

FIGURE 2-3. Township-Range system of land subdivision. See text for explanation.

The scale of a map may be expressed in other ways. A **fractional scale** ($\frac{1}{24,000}$ or 1:24,000, for example) implies that one unit on the map (the numerator) equals a certain number of units in the field (the denominator). The advantage of a fractional scale is that it is dimensionless and can be used for feet, miles, meters, or kilometers. Its disadvantages are that it cannot be used directly for measurement and, if the size of the map is changed, the fractional scale is not valid for the new map. A **verbal scale** uses words, for example, "one centimeter equals one kilometer." Occasionally, maps are described as "small scale" or "large scale." A small-scale map (say, 1:62,500) shows a large area and a large-scale map (say, 1:24,000) shows a small area on the same size map (Fig. 2-4).

Contour Lines—Vertical Distances. The vertical dimension on a map is shown by **contour lines**; these are imaginary lines that connect points of equal elevation measured from a previously established datum plane. For most maps, the datum plane has been arbitrarily chosen as sea-level and is designated as the **zero contour**. However, any map that is prepared

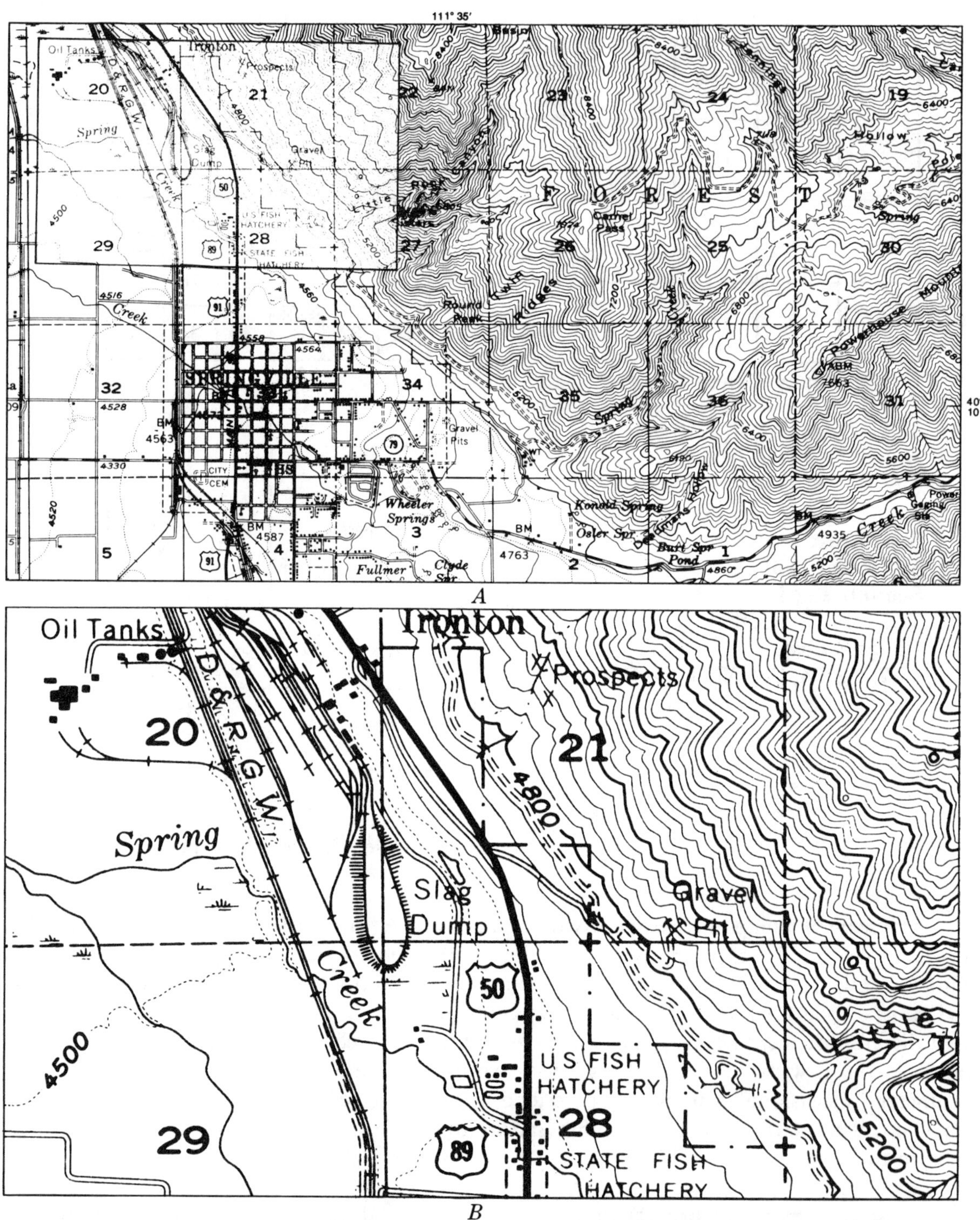

FIGURE 2-4a. The top map is a basic topographic map reduced from a scale of 1:62,600 (Provo quadrangle, Utah, 1949). 4b. The bottom map is part of the top map enlarged. The enlarged map has no greater detail and is no more accurate than the smaller-scale map (from U.S. Geological Survey Circular 721, 1976).

for a special purpose may contain a local datum other than sea-level. The vertical distance between adjacent contour lines is the **contour interval**. The interval is often 10, 20, 40, or 100 feet, although any other may be used. The interval may also be given in meters. Commonly, every fifth contour is printed with a broader or darker line and is numbered with the elevation of that line to facilitate reading the map.

A modification of the contour line appears on some maps to indicate a closed depression or omitted contour lines. This special contour line is called a **hachured contour line**. The hachures are small lines on one side of the standard contour and perpendicular to it; they point toward the center of the closed depression or toward the area where the contour lines have been omitted.

The "Rules of Contour Lines" in Table 2-2 are applicable for either reading or making topographic maps.

TABLE 2-2
RULES OF CONTOUR LINES*

1. A contour line represents a single elevation; that is, all points on the same contour line have the same elevation.
2. Contour lines do not cross other contour lines; overhanging cliffs or caves may be shown by merging contour lines or by hachure contour lines; contour lines under an overhang or cave may be shown by dashed lines.
3. Every contour line closes to define a somewhat circular area. The closure may not show on the map under consideration but may appear on adjacent maps.
4. A contour line does not separate into two or more contour lines.
5. Commonly, every fifth (or fourth) contour line is broader and its elevation shown.
6. On the same map, closely spaced contour lines indicate a relatively steep slope, widely spaced contour lines indicate a relatively gentle slope, and uniformly spaced contour lines indicate a uniform slope.
7. Contour lines cross a stream or valley at right angles to the stream or valley.
8. Where a contour line crosses a stream or valley, the contour bends to form a "V" that points upstream or up the valley.
9. Where a contour line crosses the nose of a ridge, the contour line bends to point down the ridge.
10. Where one closed contour surrounds another, the inner contour represents the higher elevation.
11. At a depression, or where contour lines are omitted because of a cliff, the contour lines are represented by hachures.
12. Where the topography slopes downhill and a standard contour is adjacent to a hachured contour, the hachured contour is one contour interval lower than the standard contour.
13. Where the topography slopes uphill and a standard contour is adjacent to a hachured contour, the two contours have the same elevation.
14. Where one closed hachured contour encloses another closed hachured contour, the inner contour is one contour interval lower.
15. Where a closed standard contour is enclosed by a hachured contour, they both have the same elevation.

*These same rules apply to bathymetric maps, which show the topography of the floors of lakes or oceans.

TOPOGRAPHIC PROFILES

Topography refers to the general shape of the land. **Relief** refers to the difference between the highest and lowest elevations in the area. Maps show the topography of an area as viewed from above. We can convert this aerial view into one that shows the topography along a particular line as seen from ground level. This is done by making a **topographic profile** along that line. The profile is like a graph with elevation plotted vertically and distance plotted horizontally. Commonly, the profile uses the same vertical and horizontal scales as the map. However, where the relief in an area is low, we can exaggerate it by doubling or tripling the vertical scale while retaining the horizontal scale.

The following steps are suggested for making a topographic profile:

1. Choose the line of the profile on the map. Note the relief along the profile line so you can estimate the values for the vertical axis (see steps 4 and 5).
2. Lay a strip of clean paper along the line of the profile. Mark the ends of the profile line on the paper using the designation of the profile line (e.g., A-A′, NE-SW).
3. Wherever a contour line intersects the strip of paper, mark a short dash at the edge of the paper. Label the index contours. It is also a good idea to label the points where streams, ridge crests, or other topographic features and certain man-made features such as roads and railroads cross the profile.
4. Lay the strip of paper on a sheet of arithmetic graph paper so that the profile line is oriented along one side of the graph paper. (The horizontal axis of the graph paper corresponds to the map distance and the vertical axis of the graph paper corresponds to the elevations. The vertical scale will not equal the map scale where vertical exaggeration is desired.)
5. Select a vertical scale (other than the map scale if appropriate). Label the vertical scale such that the highest and lowest elevations (relief) along the profile are shown. It is not necessary to start from sea level.
6. Transfer the elevations noted on the strip of paper to the graph paper. It is only necessary to plot one point for each elevation.
7. Connect the points with a smooth curve rather than straight lines. Interpolate for hill tops and valley floors.
8. Label the streams, hill tops, cultural features.
9. If you are exaggerating the vertical scale, calculate the vertical exaggeration (VE) with the equation

$$VE = \frac{\text{Representative fraction, vertical}}{\text{Representative fraction, horizontal}}$$

10. Identify the cross-section by writing on it the title, vertical scale or exaggeration, horizontal scale, name of map, date, and your name.

Figure 2-5 shows three sketches along the same profile as an example of the effect of vertical exaggeration on topographic profiles.

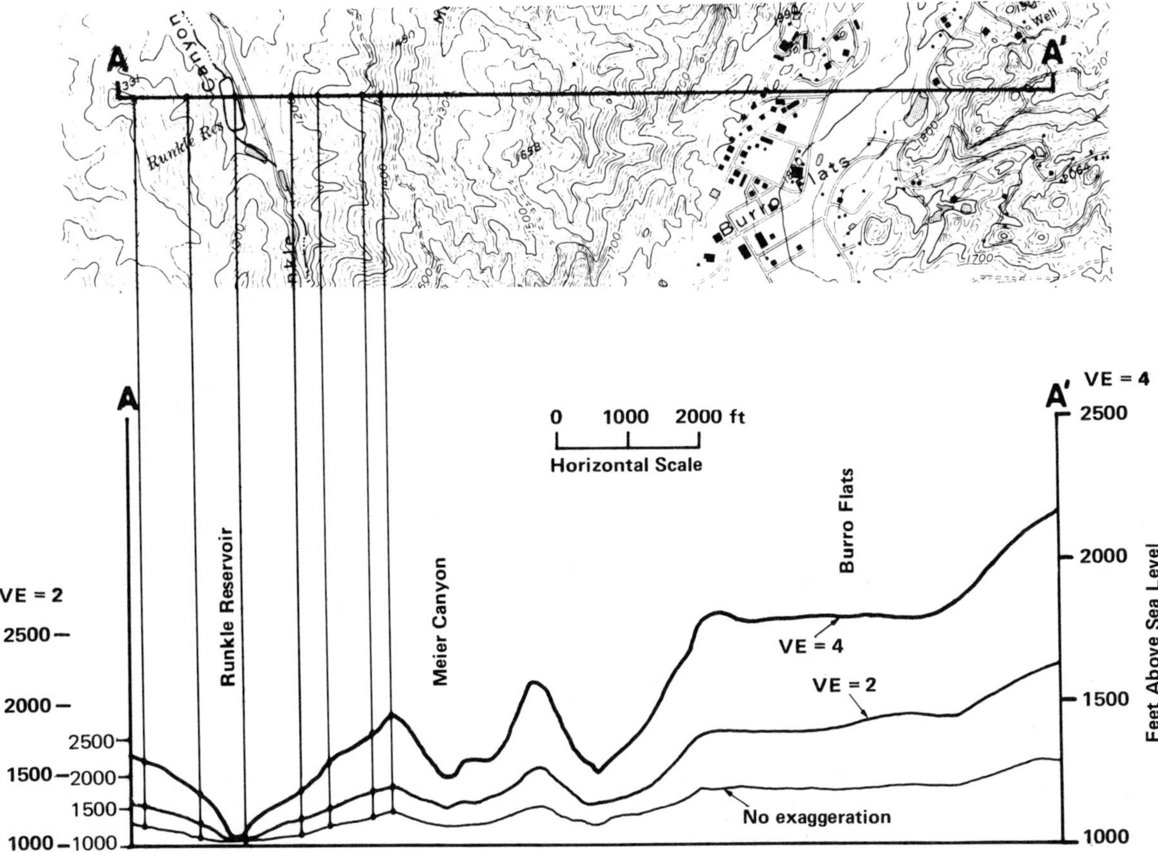

FIGURE 2-5. Topographic profiles and vertical exaggeration. A portion of the Calabasas quadrangle, U.S. Geol. Survey 7½′ topographic map (contour interval of 25 ft) has been translated into three topographic profiles, one with no exaggeration, one with vertical exaggeration (VE) of 2, and one with VE of 4. To demonstrate the technique for preparing topographic profiles, several representative points are plotted.

R9W

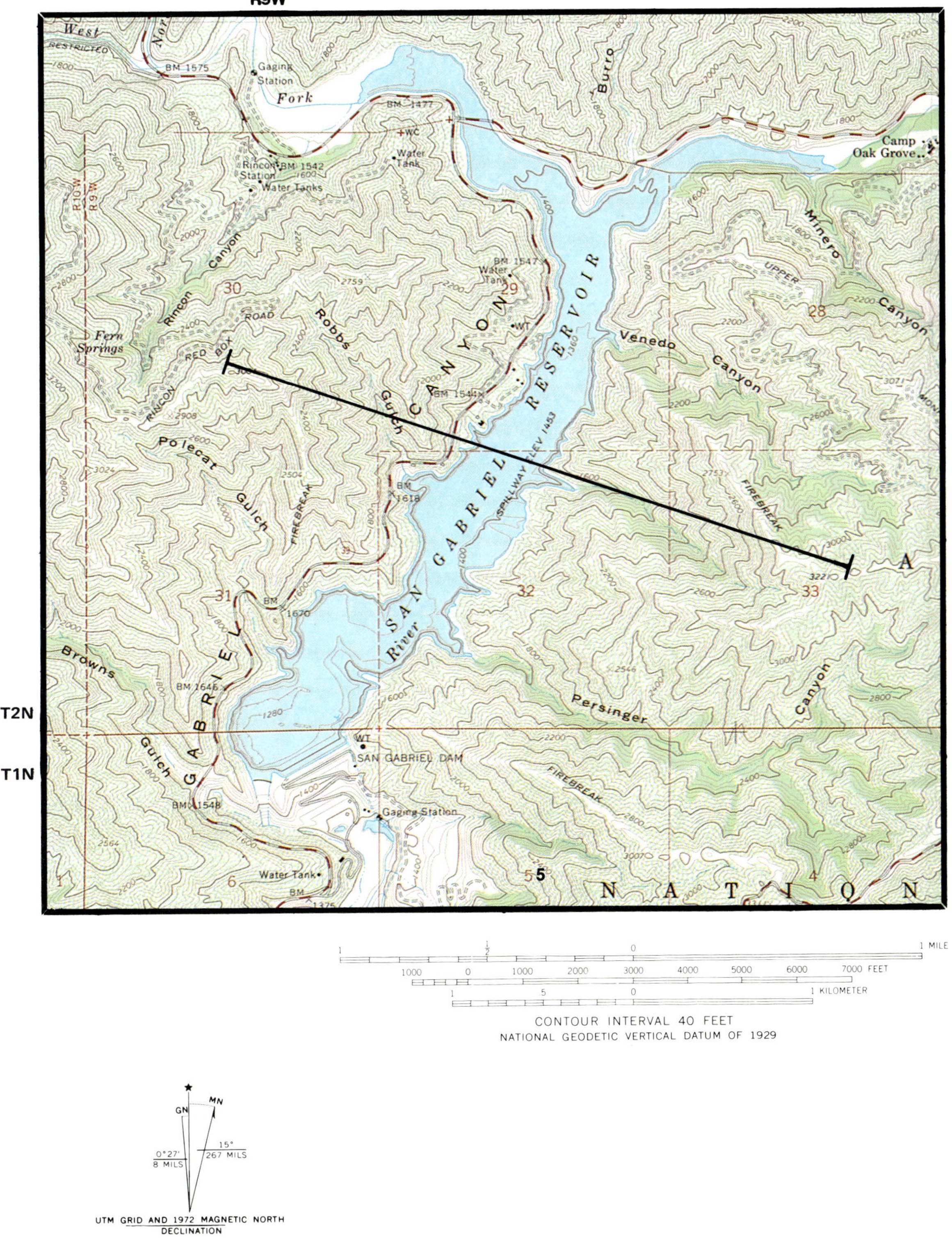

FIGURE 2-6. Portion of the Glendora quadrangle, California. (Courtesy U.S. Geological Survey.)

EXERCISE II

TOPOGRAPHIC MAPS

Name ______________________________

Refer to Figure 2-6 (p. 27), a portion of the Glendora quadrangle, California.

1. (a) What is the highest elevation in Sec 32, T2N, R9W? __________ feet.

   (b) What is the lowest elevation in Sec 32, T2N, R9W? __________ feet.

   (c) What is the relief in Section 32? __________ feet.

2. Note the blue dashed double line near the southeast corner of Sec 31, T2N, R9W. The line represents a powerplant conduit for carrying water. Follow the double line to where it emerges at the surface in Sec 5, T1N, R9W.

   (a) What is the straight line distance of the conduit? __________ miles.

   (b) What is the true distance along the conduit? __________ miles.

   (c) What is the gradient of the conduit? __________ feet/mile.

3. (a) What is the greatest depth of water in the San Gabriel Reservoir? __________ feet.

   (b) The surface area of water in the San Gabriel Reservoir can be approximated by ignoring irregularities along the shoreline and considering the surface area to be composed of a number of squares and rectangles. Overlay a clean sheet of paper on the lake and sketch in rectangles such that the lake is almost completely enclosed. Then simply sum the area of each rectangle to get the total area. You will probably have to ignore the extreme ends of the "arms" at the north end of the reservoir.

   Surface area ______________ square miles; __________ square kilometers.

4. On the grid below, draw a topographic profile from elevation 3221 (near the center of Sec 33, T2N, R9W) to elevation 3001 (south of the center of Sec 30, T2N, R9W). Show the topography across the floor of the reservoir to true scale horizontally and vertically. Draw the same section with a vertical exaggeration of 4 (4X). Show the elevation of the water in the reservoir.

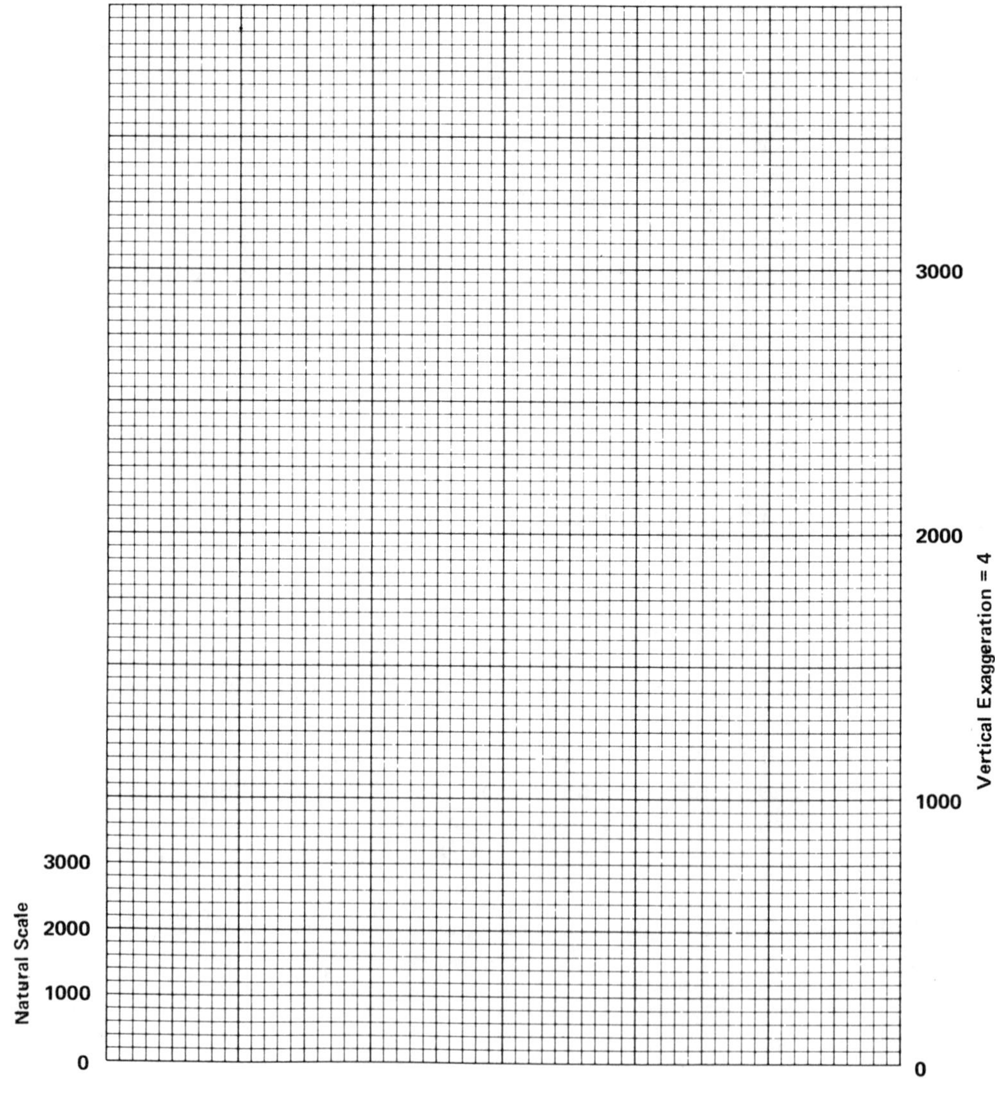

Chapter 3
Geologic Maps

Geologic maps show the distribution of rock types as well as their stratigraphic and structural rock relationships. Geologic maps are two-dimensional representations of three-dimensional reality. Commonly, the geologic information is superposed on a topographic map of the same scale. Great care is used to plot the geologic information accurately on the topographic map in order to maintain the correct field relationships.

Because the geologic data are superposed on a topographic map, the information on the topographic map and in its margins also appears on the geologic map. The contour lines, scale, directions, cultural features, and so forth are all present. The additional geologic information appears as colors and/or patterns that represent the rock formations and/or rock types. Special geologic symbols provide details of the geology. The three-dimensional interpretation from the two-dimensional map is obtained by using the relation of the rock units among themselves and with respect to the topographic map, as well as from the information supplied by the geologic symbols.

TYPES OF GEOLOGIC MAPS

There are two fundamental types of geologic maps: surface geologic maps and subsurface geologic maps. Geologic maps that emphasize specific geologic hazards for engineering or environmental analyses are becoming more common. These maps may also be either surface or subsurface.

Surface geologic maps show the distribution of rock types and the relation among them at the surface of the Earth. The maps also show the attitudes (horizontal, tilted) of the rocks at the surface by the use of special symbols. The information from which these maps are made comes from field observations, interpretation of aerial photos, and other sources of data. The relation among the rocks includes their relative ages, whether the rocks are faulted or folded, and so forth. In general, weathered soils, alluvium, talus, glacial deposits, and similar earth materials are not shown on geologic maps unless these deposits are relatively thick. Special maps, called **surficial geologic maps**, show the distribution of such deposits.

Subsurface geologic maps show the distribution of a particular rock type at depth (e.g., coal), the thickness of the rock type, the configuration of the top of the unit, or areas that have been mined out. The information from which these maps are made comes from logs from drill holes, geophysical data, or geologic mapping in mines.

GEOLOGIC TERMS DEFINED

Strike. The orientation of any plane in space may be described by the strike and dip of that plane (Fig. 3-1). **Strike** is the trace of the inclined plane on the horizontal plane. It is referred to north and, in the case shown, the strike is north 30° west (N30W) and its azimuth is 330° (330° clockwise from north).

Dip. The vertical angle that the inclined plane makes with the horizontal plane is its **dip**. Dip is taken perpendicular to strike because any other angle will give an "apparent dip" which is less than the true dip. A plane that strikes northwest can dip either southwest or northeast. Thus when describing the attitude or orientation of a plane in space the strike and dip directions must be specified; thus, N30W, 60° SW describes the plane in Figure 3-1. The true dip of that plane is 60 degrees to the southwest.

Anticline. An anticline is a convex upward fold of layered rocks (Figs. 3-2 and 1-4). The oldest rocks are in the center of the fold.

Syncline. A syncline is a concave upward fold in layered rocks (Figs. 3-2 and 1-4). The youngest rocks are in the center of the fold.

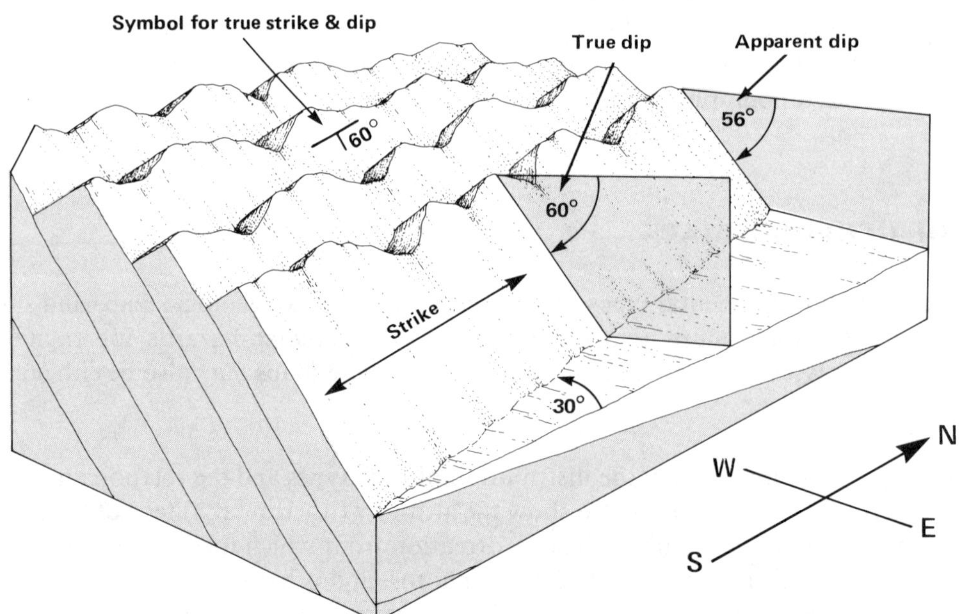

FIGURE 3-1. Block diagram showing dip and strike. Strike is 30° west of north (N30W) and dip is 60° to the southwest (60°SW). Full notation: N30W, 60°SW.

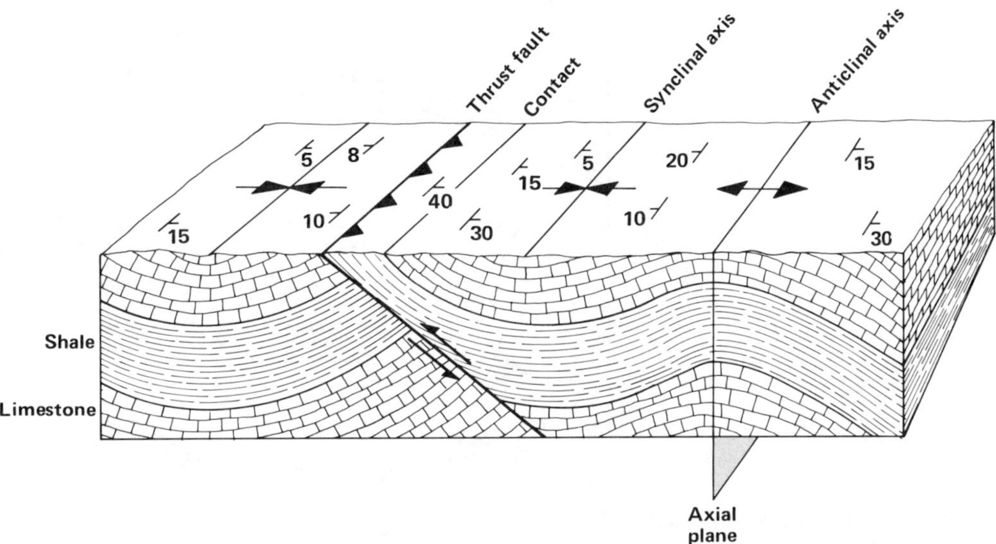

FIGURE 3-2. Block diagram showing geologic cross-section of the folded Appalachian Mountains in Tennessee. For clarity, geologic structure symbols are shown on the map surface without topography or rock formations.

Axial Surface. A surface formed by connecting lines of maximum curvature (hinge lines) of a fold is called an **axial surface**. When the axial surface is planar (not curved) it is an **axial plane** (Figs. 3-2 and 3-3). When the limbs of a fold dip equally in opposite directions the axial plane is vertical and the fold is symmetrical (Fig. 3-3a). When one limb of the fold dips more steeply than the opposite limb the axial plane is inclined and the fold is asymmetrical (Fig. 3-3b).

Fault. A fracture in the Earth's crust along which there has been some movement (displacement) is known as a **fault**. A fault that dips at a low angle and in which the upper block of rock has moved up relative to the underlying rock is known as a **thrust fault**. See Chapter 6 for illustrations of the geometry of faults.

COLORS, PATTERNS, AND SYMBOLS

Maps prepared by the U.S. Geological Survey commonly show rock formations in different colors. Although the color scheme used by the U.S. Geological Survey is standard within that organization, it is not universal. Even though the color provides information on the relative ages of the rocks, it does not always provide information on rock type. However, a description of the rocks is given in the legend of the map. The legend appears in the margin.

Patterns may be used on the map to distinguish rock types or formations (Table 3-1). The patterns may be in color or in black and white and occasionally a pattern is superposed on the color to differentiate rock units within the formation. In this way, for instance, a clay layer within a sandstone formation, which acts as a slip plane in landslides, can be easily identified on a map. Patterns are also used in geologic cross-sections to convey the relative structural complexities.

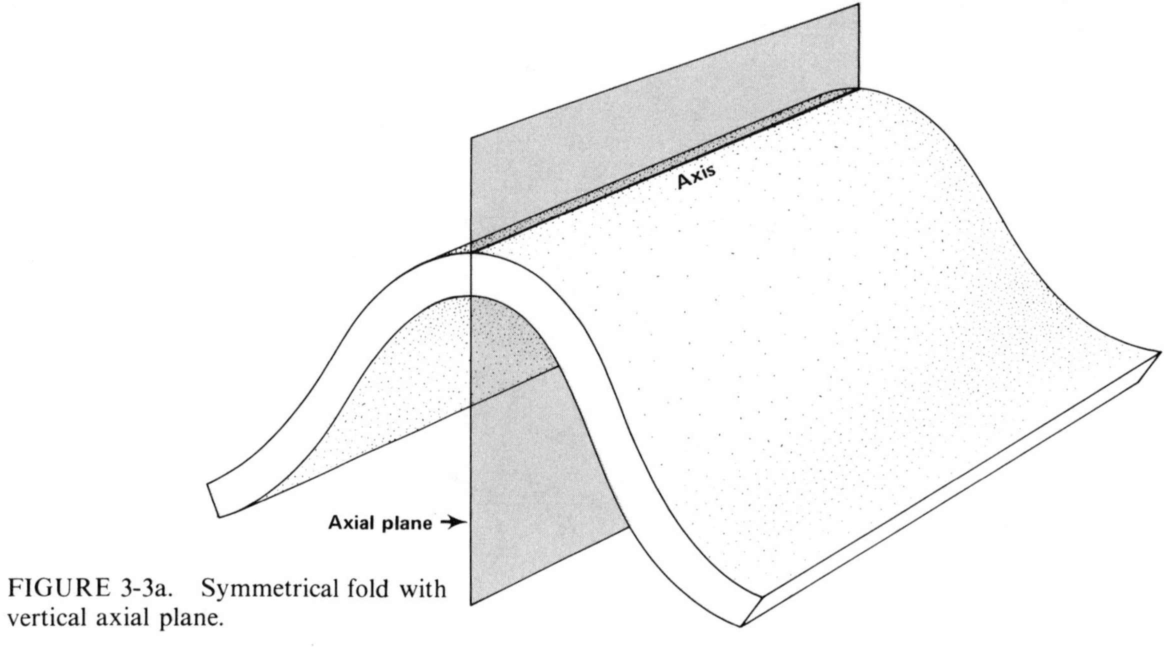

FIGURE 3-3a. Symmetrical fold with vertical axial plane.

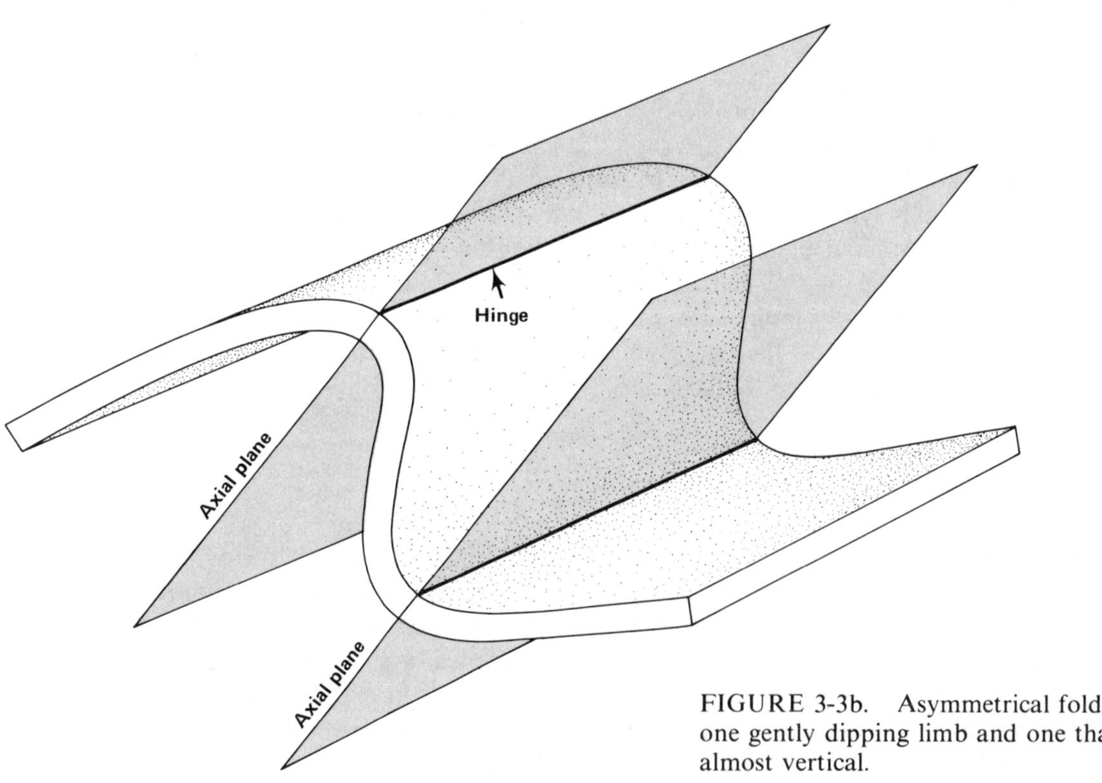

FIGURE 3-3b. Asymmetrical fold with one gently dipping limb and one that is almost vertical.

TABLE 3-1
GEOLOGIC SYMBOLS FOR ROCK TYPES

Igneous	Sedimentary	Metamorphic
 Granite	 Sandstone	 Quartzite
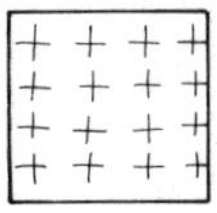 Diorite	 Shale	 Slate
 Gabbro	 Siltstone	 Schist
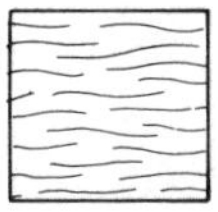 Rhyolite	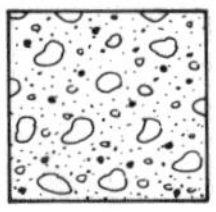 Conglomerate	 Gneiss
 Basalt	 Limestone	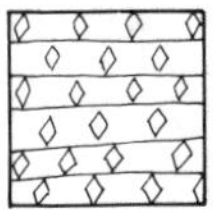 Marble
	Dolomite	

GEOLOGIC MAP
OF THE
HAYWARD QUADRANGLE, CALIFORNIA
By
G. D. Robinson
1956

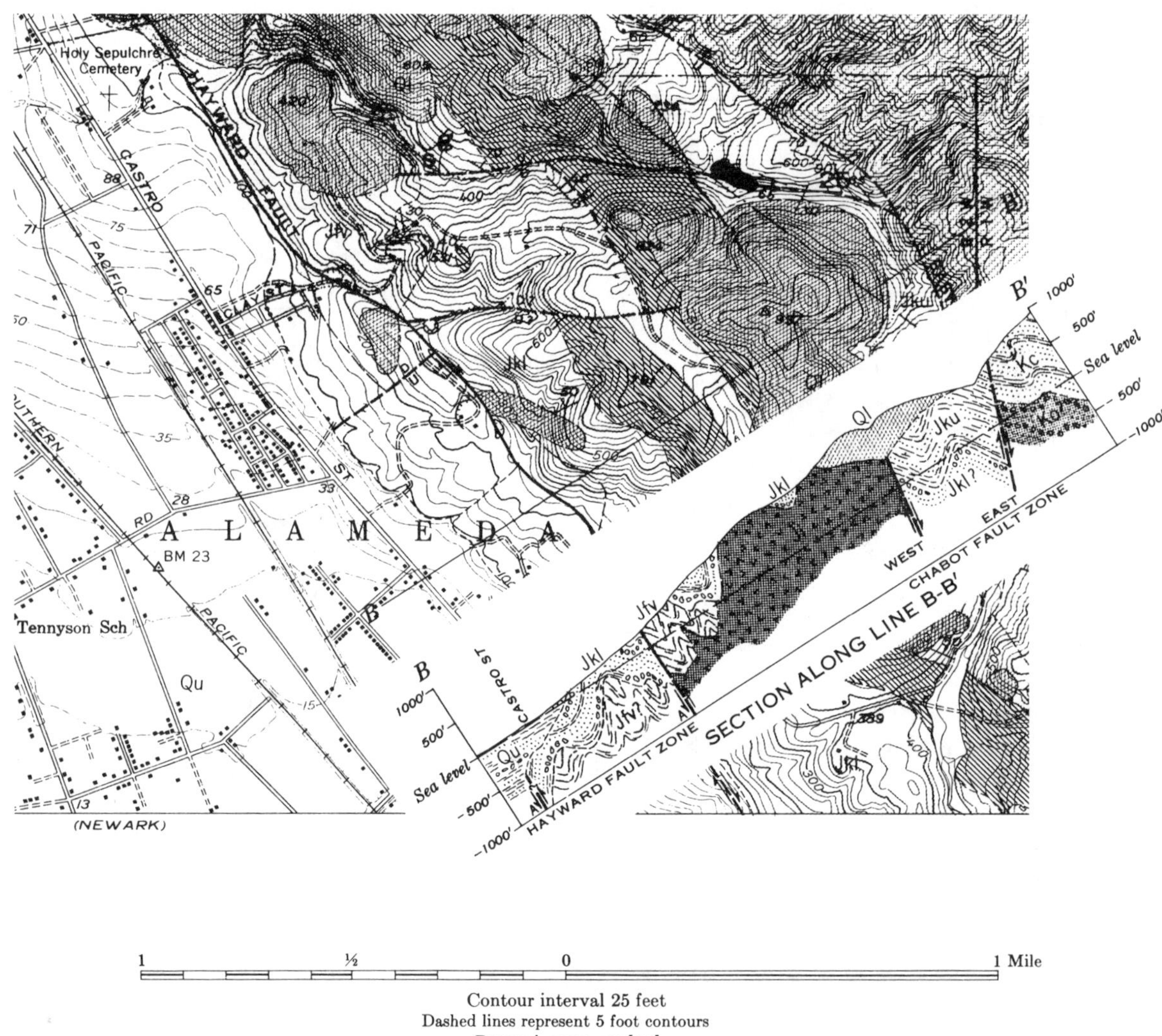

FIGURE 3-4. Geologic cross-section, Hayward quadrangle, California. Rocks of Jurassic age-Franciscan Group (Jfv) overlain by Jurassic Knoxville Formation (Jku, Jkl) and Cretaceous Oakland Conglomerate (Ko) and Chico Formation (Kc). Very young Quaternary rocks and sediments are designated (Ql) and (Qu). (Courtesy U.S. Geol. Survey.)

Special symbols have been developed that provide details of the geology. These symbols are a shorthand notation of the geologic formations at a location, such as folds and faults. For example, the attitude or inclination of a rock in an outcrop is shown by a strike and dip symbol (Fig. 3-1). There are also symbols for other common geological structures (See Appendix D). Symbols have also been developed to designate the age of the rocks. In the legend of the map, the ages are given using capital letters and they are listed in descending order, with the youngest rock in the map area written at the top and the oldest at the bottom. The formation's designation appears as a lower case letter (or letters) following the age designation. Whereas the age designation is standard, the formational designation is not.

LEGEND OR EXPLANATION

The legend (or explanation) includes all the colors, patterns, and symbols used on the geologic map and defines their specific application. In addition, the legend also contains the ages of rocks in the area and a brief description of the rock types within each formation. Engineering and environmental geologic maps may contain information on the engineering characteristics of the rocks. The name of the geologist who mapped the area as well as the year the area was mapped also appear in the margin.

GEOLOGIC CROSS-SECTIONS

To help convey three dimensions from a two-dimensional geologic map, we can make a **geologic cross-section**. We use the same techniques as in preparing a topographic profile (see Ch. 2). The locations of the rock types and their attitudes, contacts between rock types, location and attitudes of faults, and similar information shown on the map are projected onto the cross-section. The geologic cross-section also contains the topographic profile. Vertical scale is **not** exaggerated in either the profile or the cross section (Fig. 3-4).

We can infer subsurface conditions (such as rock types and their attitudes) from the surface geology with the aid of the special symbols. Such a geologic cross-section is an extrapolation from what is known about surface conditions and can never be as accurate. It requires experience with geology in general and a knowledge of the local geology in particular.

EXERCISE III

GEOLOGIC MAPS

Name ______________________________

Geologic mapping and analysis of folds and faults provide important information about the geology at depth. For example, petroleum and natural gas migrate from source rocks and collect in reservoir rocks, such as sandstones, at the tops of anticlines or domes (see Chapter 6). Such knowledge of rock structures helps geologists to locate new oil fields and new producing horizons in old oil fields. Another example of information at depth is where a fault cuts an ore-bearing vein or sedimentary stratum. Determination of the direction and amount of offset on the fault can aid in locating the vein on the opposite side of the fault (see Fig. 6-2 for relative motion on faults).

The photo of Figure 3-5 and the geologic map of Figure 3-6 show a small part of the Rocky Mountain overthrust belt in Lincoln County, Wyoming. The rocks are of Mesozoic and Cenozoic age (see Appendix C) and most of the faults are thrust faults dipping gently to the west. Folding and faulting took place during late Cretaceous time and then younger Tertiary deposits were laid down on top of the folded rocks with little subsequent deformation of the younger strata (see Fig. 3-6, cross-section). Coal is a major resource in the area, with more than a million tons of bituminous and subbituminous coal stripped annually from the Frontier and Adaville Formations. The town of Kemmerer was founded in the late 19th century to exploit the nearby coal deposits of Cretaceous age. Oil and gas are now of great interest because of recent major discoveries in folded rocks associated with intense deformation of the overthrust belt in Cretaceous time. The area is also known for its phosphate and oil shale reserves.

1. Complete Figure 3-6's geologic cross-section B—B′ of the folded Cretaceous and flat-lying Tertiary rocks in the Kemmerer quadrangle, Wyoming. Use the dips and strikes shown on the map nearest the cross-section line to determine the subsurface structure. To do this place the edge of a clean sheet of paper along the section line and mark the contacts of the different formations and indicate the direction and amount of dip where shown on the map. Put the symbol for the age and name of each unit on the section. Work slowly and carefully and you should complete the section on the first try without any problems.

2. Based upon the completed cross-section, are the axial planes of the folds vertical or do they have a gentle dip? ________________ Are the folds symmetrical or asymmetrical?

   ____

3. Assume for the moment that the sedimentary unit, Kg, contains oil (in reality it might). On the cross-section, locate a vertical drill hole to intersect the hinge line of the folded sandstone unit, Kg. You can see that the drill hole is not located at the intersection of the ground surface with the axial plane of the fold, where the hinge is indicated. From this, you should be able to see the advantages of being able to construct accurate geologic cross-sections from surface and drill-hole data.

4. Where else might oil have accumulated in the Gannett Formation (Kg) along the line of section? (See Chapter 13, Mineral Resources and Fossil Fuels, for a diagram of a similar geologic situation.) ______________________________

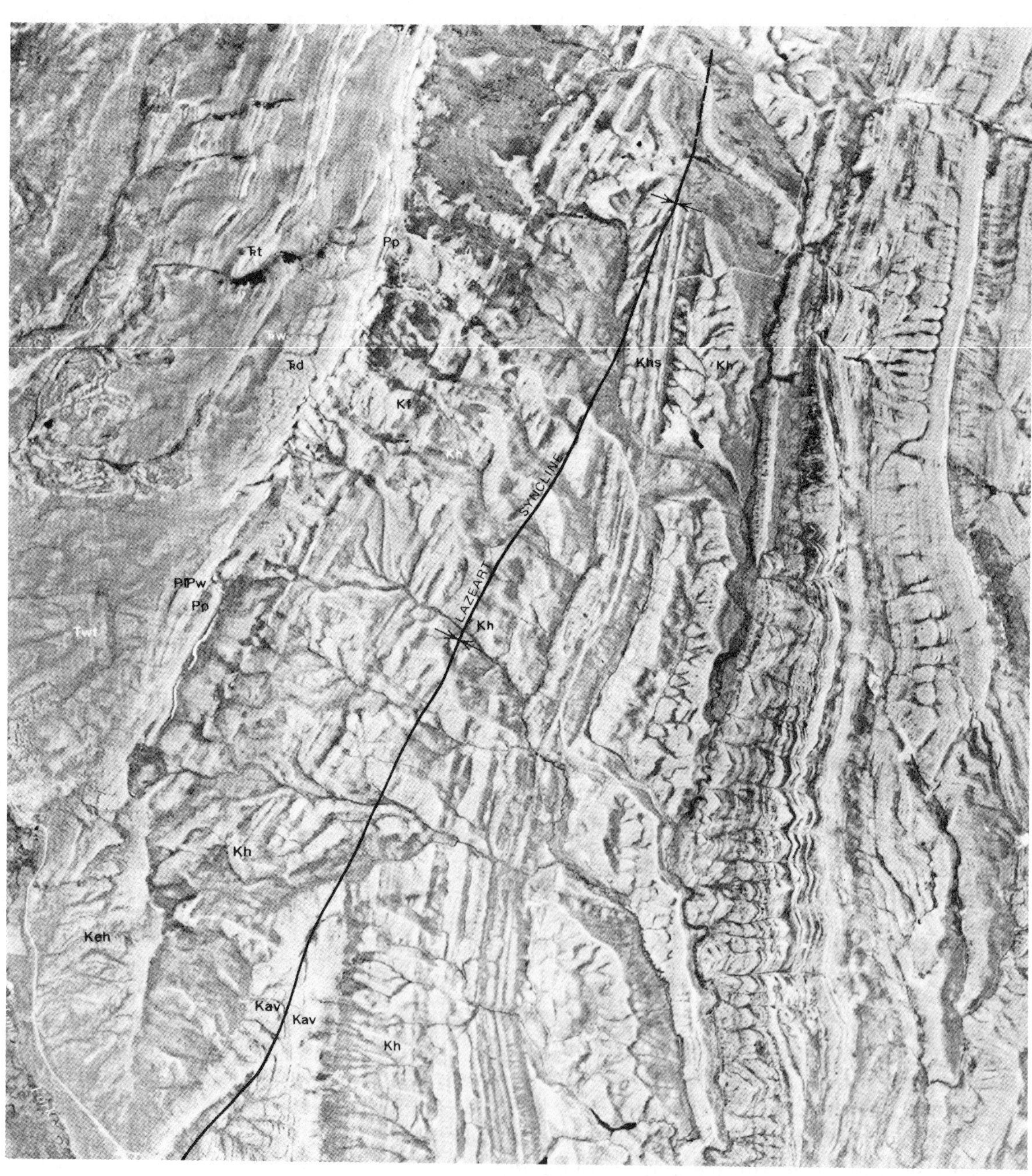

FIGURE 3-5. Northeast part of Kemmerer quadrangle, Wyoming. (Courtesy U.S. Geol. Survey.)

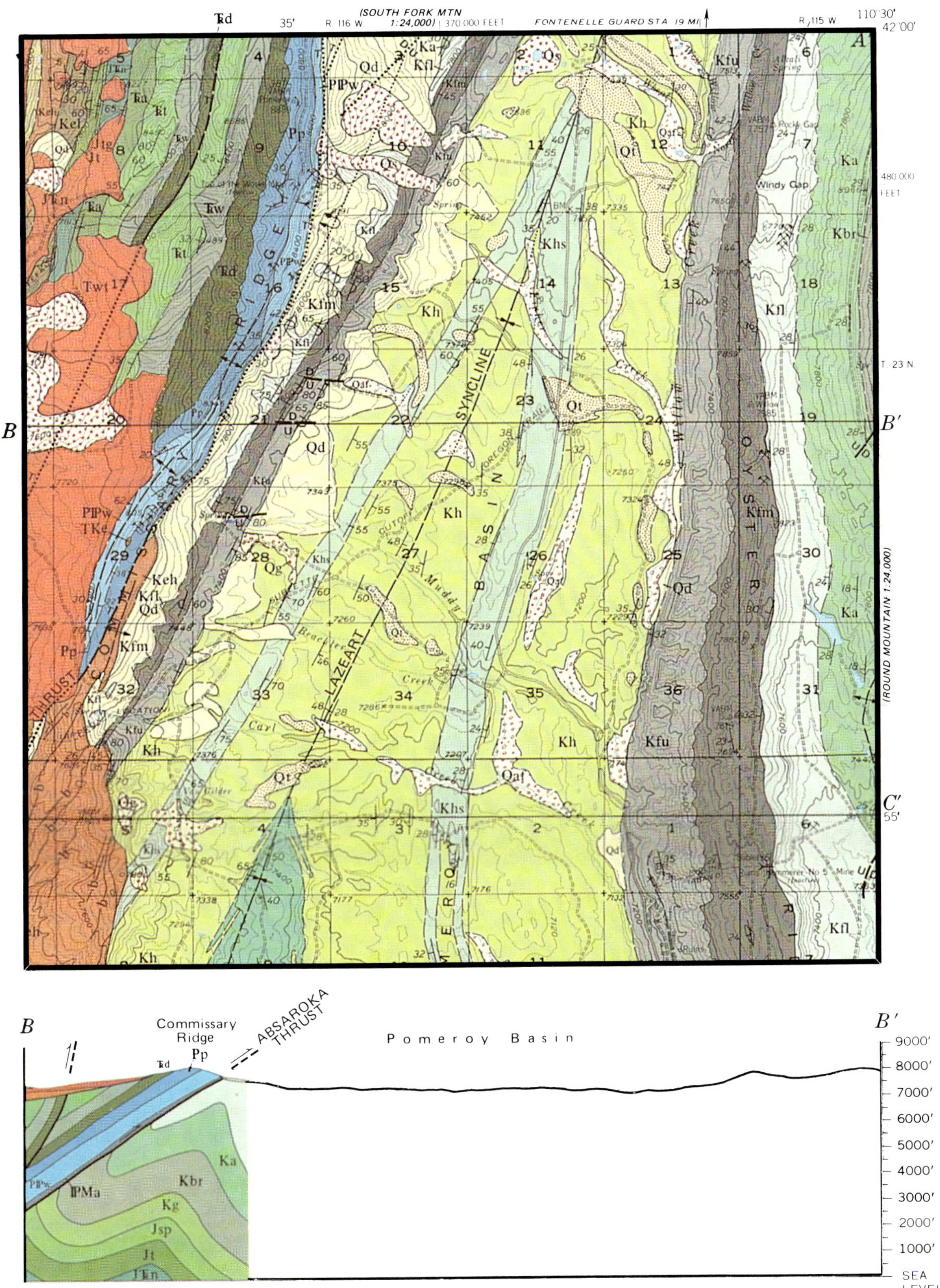

FIGURE 3-6. Geologic map and cross-section of the Sage and Kemmerer quadrangles, Wyoming. (Courtesy U.S. Geol. Survey.)

Chapter 4

Earthquakes and Their Location

Earthquakes are measurable vibrations of the Earth's crust caused by abrupt transient movements along faults. These movements release stored strain energy and result in rock displacement, ground shaking, and heat loss. The theory behind the generation of earthquakes by movements on faults is called the **elastic rebound theory**. The theory proposes that rocks are subjected to stresses and that elastic strain accumulates in them. Eventually the strength of the earth materials is exceeded, rupture occurs, and **seismic waves** are generated (Fig. 4-1). If the rock displacement is sufficiently large and close to the surface, surface rupture of the ground may occur.

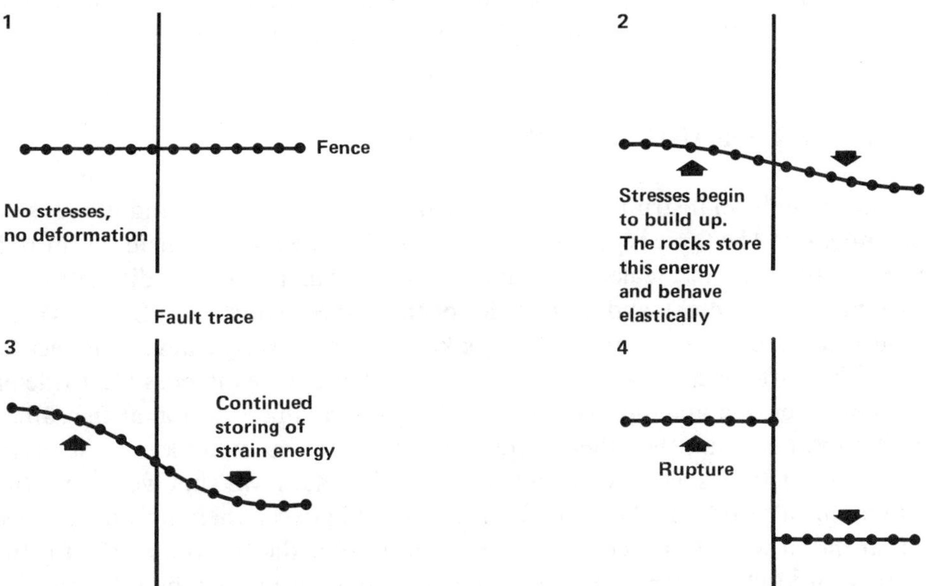

FIGURE 4-1. Accumulation of strain energy according to the elastic rebound theory. Seen from overhead, a fence crosses a fault trace. As stress accumulates, the ground and fence deform. Stress eventually exceeds the strength of the rock and rupture occurs with an accompanying earthquake.

TABLE 4-1
PROPERTIES OF SEISMIC WAVES

Body Waves	P-waves	Primary waves move with the greatest velocity and are the first to arrive at the surface. They are compressional waves with particle displacement in the direction of propagation (Fig. 4-2).
	S-waves	Secondary waves are slower than P-waves and arrive at the surface later. They are shear waves with particle displacement perpendicular to the direction of propagation (Fig. 4-3). They can cause damage to man-made objects.
Surface Waves	R-waves	Slowest waves; travel only at the Earth's surface. They are the most destructive and exhibit retrograde and elliptical particle motion (Fig. 4-4).

The vibrations that we call an earthquake spread spherically from the **hypocenter** (also known as the **focus**), the point inside the Earth where the earthquake originates. The point on the surface of the Earth directly over the focus is called the **epicenter**. Earthquake hypocenters have been located to depths as great as 700 km. Below this depth the Earth is apparently plastic and may deform continuously rather than in a sudden burst of released energy. The three types of elastic waves produced by an earthquake are described in Table 4-1.

These seismic waves constitute the shaking we feel during a quake and produce the observed damage. The **P-** and **S-waves** move through the Earth and are called body waves, but they propagate with different velocities and different motions (Figs. 4-2 and 4-3). The surface wave, also called a **Rayleigh** or **R-wave**, is the slowest. It sets the rock or soil particles moving in an elliptical path. At the top of the ellipse, the particle is moving in a direction opposite to that in which the wave is propagating; this movement is called "retrograde" (Fig. 4-4). The surface wave produces the rolling motion felt during an earthquake.

LOCATING THE EARTHQUAKE EPICENTER

A **seismograph** is an instrument used to record earthquake waves and it does so on a chart called a **seismogram**. The three types of waves can be identified on the seismogram and scientists make interpretations and calculations based upon arrival times and differences in time of arrival (distance to epicenter), and amplitudes of the waves. The P- and S-waves are generated simultaneously at the hypocenter of the quake. The P-wave, however, travels faster than the S-wave. This difference in velocities, or travel times, makes it possible to determine the distance to epicenter. An analogy would be two trains leaving a station at the same time, one traveling 60 miles per hour, the other 30 miles per hour. If you didn't know the origin of these trains, but you knew their speeds, you could calculate how far away they were when they started. To do this you must also know how much time elapsed between their arrival at the same point. Let's say that the slow train passed your house 1 hour after the fast train. What is the distance to the point of origin? (answer: 60 miles) Similarly, since we know the velocity of the P- and S-waves we can determine the distance to epicenter by knowing the difference in their arrival times at a seismograph station.

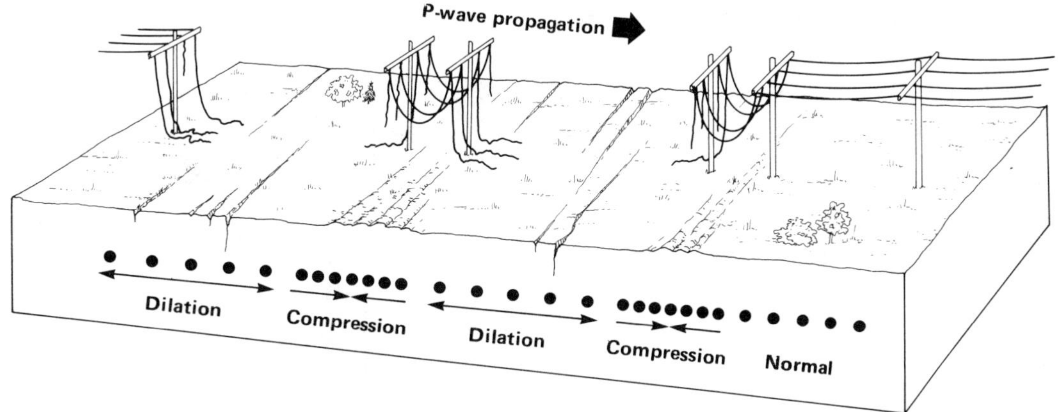

FIGURE 4-2. Ground motion during passage of a P-wave.

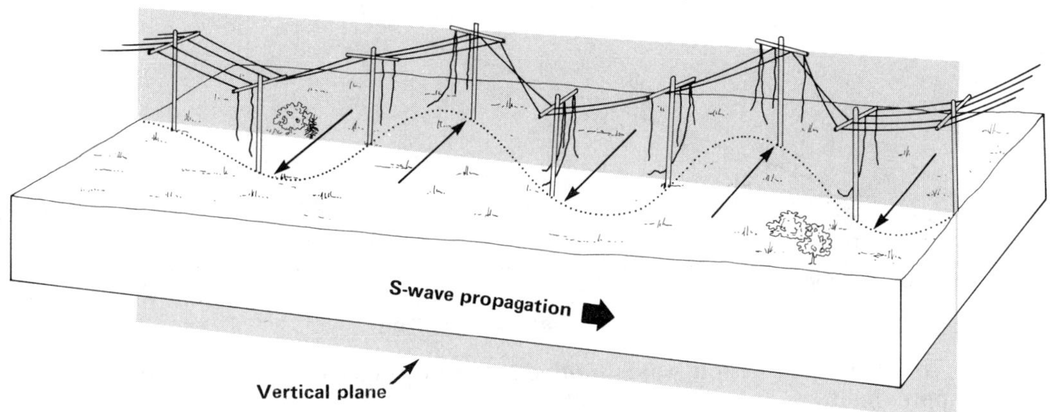

FIGURE 4-3. Ground motion during passage of an S-wave.

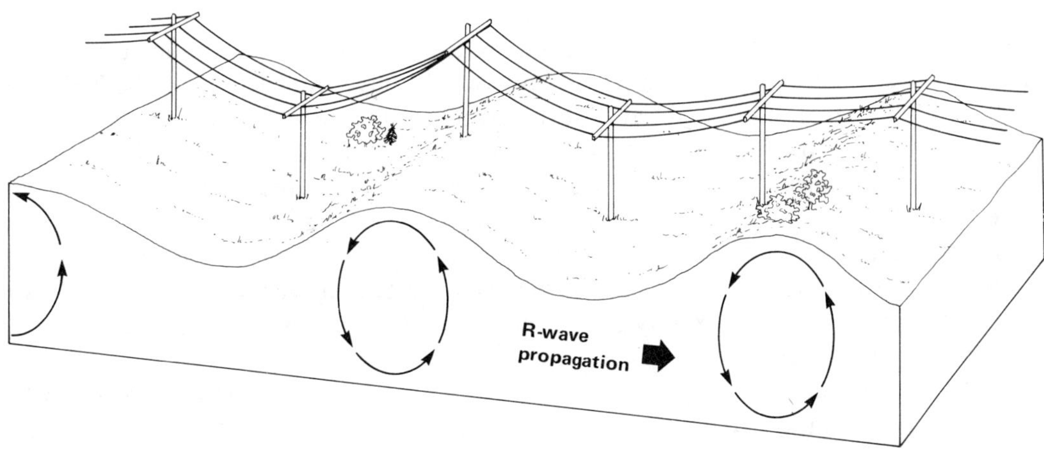

FIGURE 4-4. Ground motion during passage of a surface wave, or R-wave. Note elliptical pattern of solid particles and that the particle at top moves in direction opposite to wave propagation.

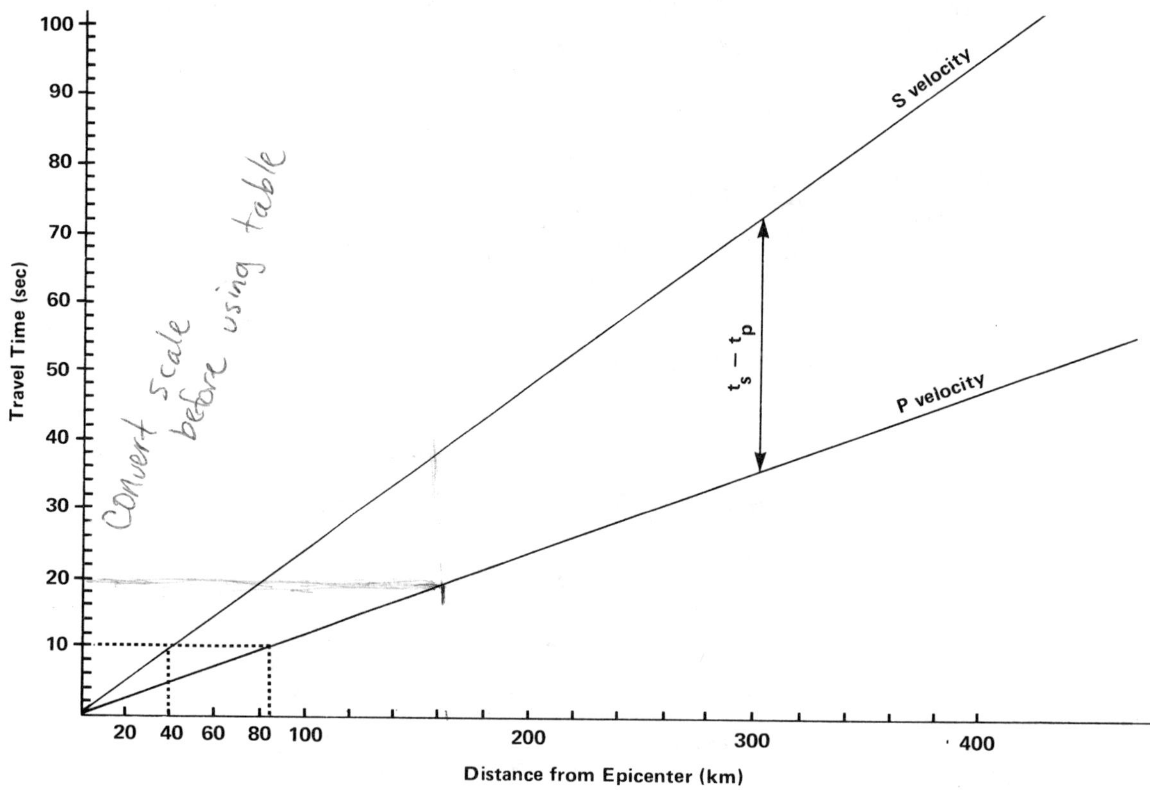

FIGURE 4-5. Travel time of seismic waves plotted against distance from epicenter ($t_S - t_P$). Seismic velocities are approximate.

Figure 4-5 is a graph of travel time versus distance for the P- and S-waves, the plotting of which results in curves that represent S- or P-wave velocities. That is, the ordinate shows the travel time in seconds and the abscissa shows the distance travelled in kilometers. Thus we see that for a travel time of 10 seconds the P-wave would have travelled about 82 km and the S-wave about 40 km. It should be noted that the graph is an approximation of more precisely measured travel times. Now, if you know the difference in the arrival times of the two waves you can determine the distance to epicenter simply by scaling off on the graph. In the example shown the time difference ($t_S - t_P$)* measured from the seismogram at a given station was 35 seconds. Along the ordinate, the length of 35 seconds was scaled off on a piece of paper. This length was then shifted along the space between the two curves for the S- and P-waves until the gap between them was exactly 35 seconds. The distance to epicenter was then found along the abscissa to be 300 km.

To find the epicenter or hypocenter of the quake you need more data than the distance from just one seismic station (Why?). In fact you need distances from at least three stations to reasonably locate the epicenter (Again, why?).

---

*t_S and t_P are the travel times, in seconds, of the “S” and “P” waves.

EXERCISE IV

DETERMINATION OF EARTHQUAKE EPICENTERS

Name ______________________________

The purpose of this exercise is to understand the procedure for locating the epicenter of an earthquake and for determining the time of the quake. You are given one set of 8 seismograms which record an earthquake that occurred on December 19, 1974 (Fig. 4-6). A map is also provided which shows the location of the seismic stations in southern California that recorded this earthquake (Fig. 4-7). Abbreviated station codes are as follows.

SYP	Santa Ynez Peak	GLA	Glamis	PYR	Pyramid
MWC	Mt. Wilson	PLM	Palomar	CSP	Cedar Spring
ISA	Isabella	GSC	Goldstone		

Minute marks are shown on the seismograms and a time scale is located between the seismograms ISA and GLA.

(A) Mark an arrow on each seismogram to indicate the arrival of each P- and S-wave. (S-wave starts at the beginning of a group of large oscillations.) For some stations closer to the epicenter (MWC, PYR, and CSP), the S-wave arrives before the P-wave oscillations have subsided. This situation will make it impossible to determine the exact S-wave arrival time.

(B) Make a list of arrival times of P-waves and S-waves (in hours, minutes, seconds) on the data sheet on p. 51. Also record the time lag ($t_S - t_P$) for each station.

(Note: for those stations where it is impossible to determine both P-waves and S-waves, you will not be able to determine the time lag.)

(C) Locate the epicenter of the earthquake. To do this you must first determine the distance from the focus of the earthquake to each station using the ($t_S - t_P$) graph (Fig. 4-5). List these distances on your data sheet. Next, plot this information on Fig. 4-7. From each station on the outline map for which you have data draw a circle whose radius is equal to the distance to the focus of the quake. This circle is the locus of points a fixed distance from the station. The earthquake epicenter will be located at the point where all of the circles intersect. Unfortunately, inaccuracies are introduced in reading the seismograms and graphs, in plotting a spherical surface on a flat map, and in calculating distances to the epicenter rather than to the focus. Therefore, the circles do not generally intersect precisely at one point. Instead of a point you will see an area. The epicenter is located at the center of this area.

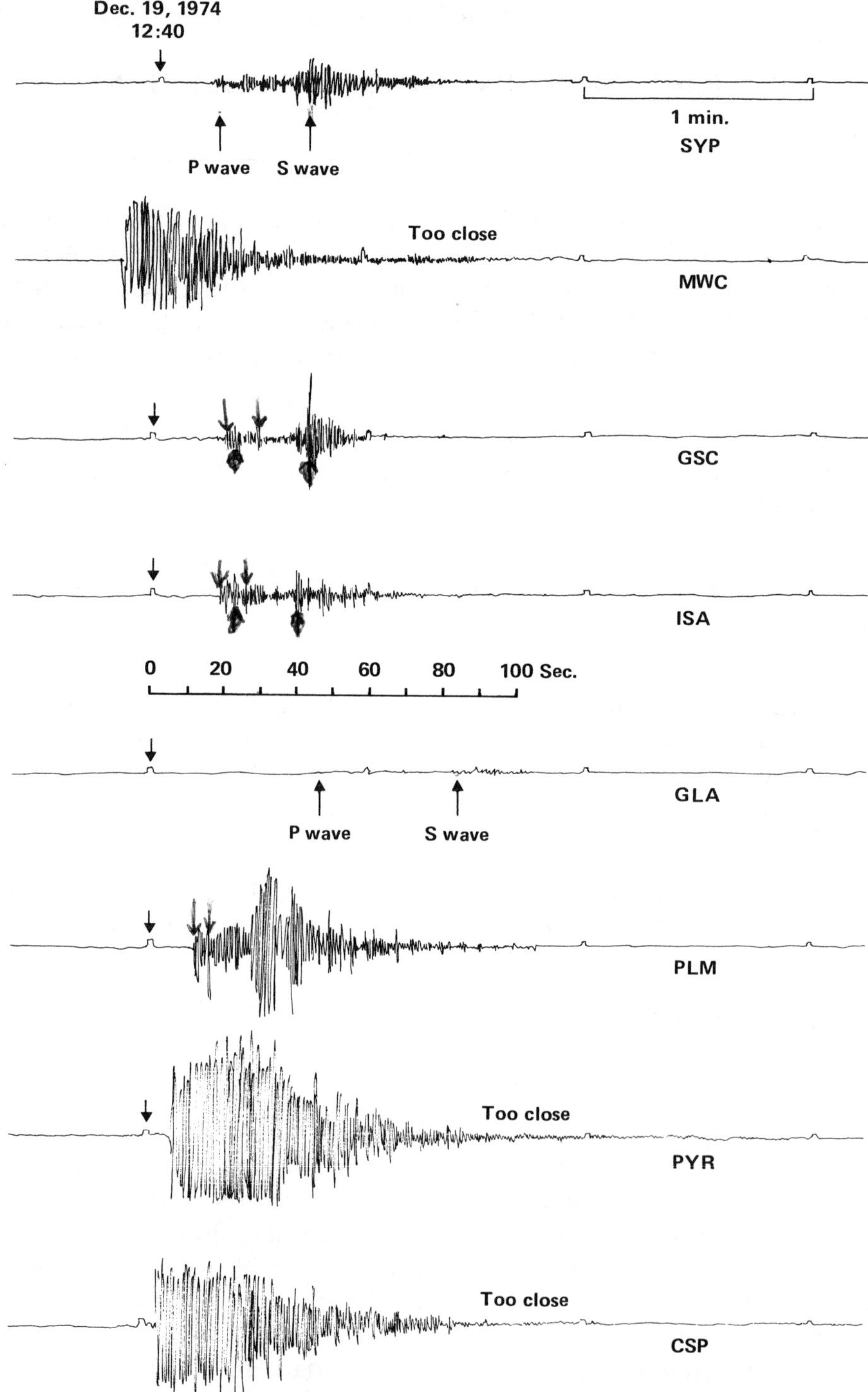

FIGURE 4-6. Seismograms for the earthquake of December 19, 1974, at eight southern California stations.

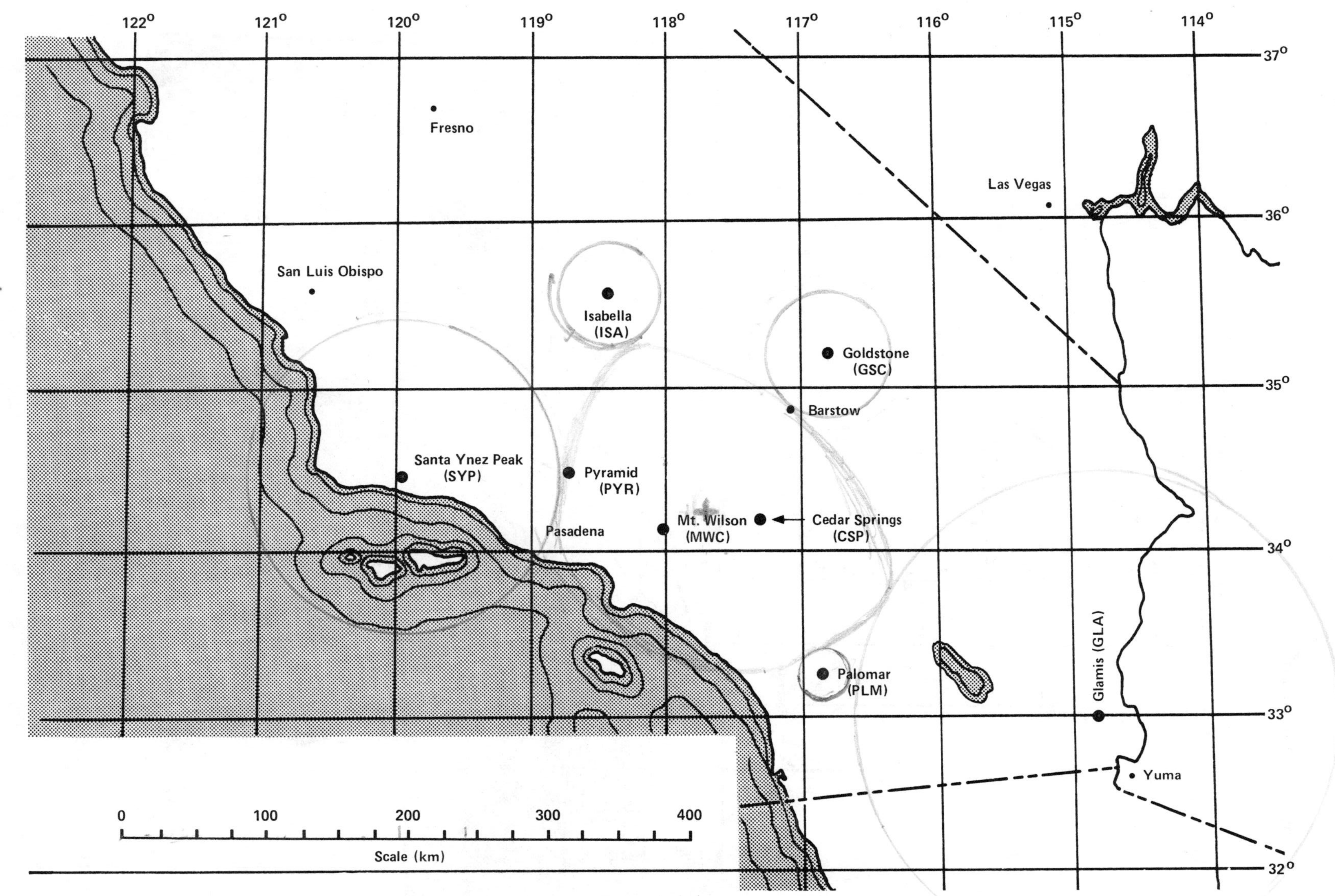

FIGURE 4-7. Seismograph station locations for earthquake of December 19, 1974.

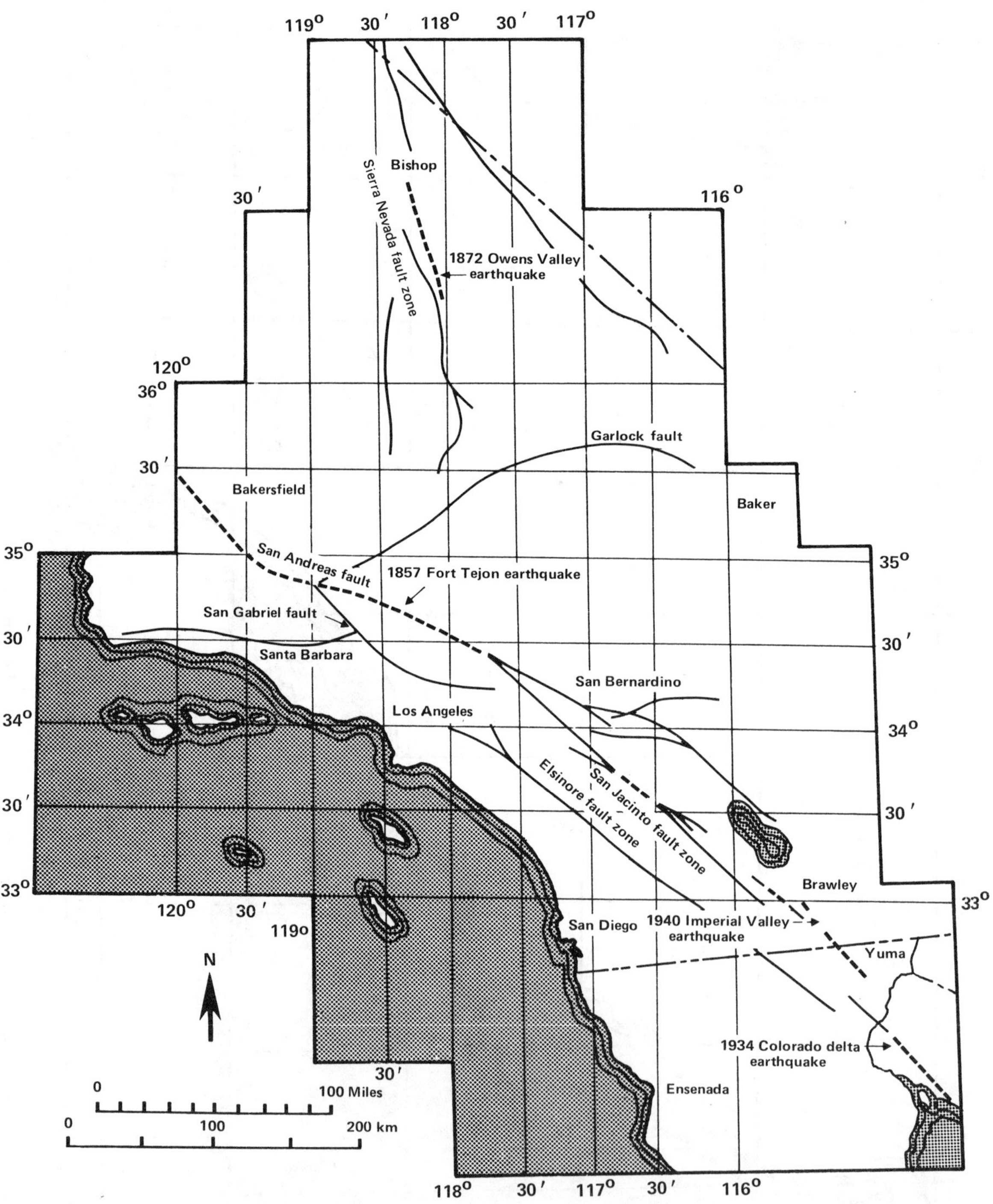

FIGURE 4-8. Major active faults in southern California. Significant earthquakes are shown.

redue

DATA SHEET

Station	Arrival Time		$t_S - t_P$	Epicentral Distance	
	P-Wave	S-Wave			
SYP	12:40 19s.	12:40 42s	23	110km.	185
MWC	too close		—	—	
GSC	20s.	30s.	10 20	40km.	163
ISA	19s.	26s.	7	30km.	163
GLA	47s	84s.	37	170km.	313
PLM	12s	16s.	4	20km.	138
PYR	too close		—	—	
CSP	too close		—	—	

1. What is the location of the earthquake epicenter, in latitude and longitude coordinates?

   117.7° lat. + 32.2° long 34°08'N 118°05'W

2. What is the approximate time of origin of the quake? (Use the $t_S - t_P$ graph, which relates time and distance.) 12:39 54sec.

3. Look at the California Fault Map (Fig. 4-8). What fault might have produced this event?

   Elsinore or San Gabriel/Santa Barbara

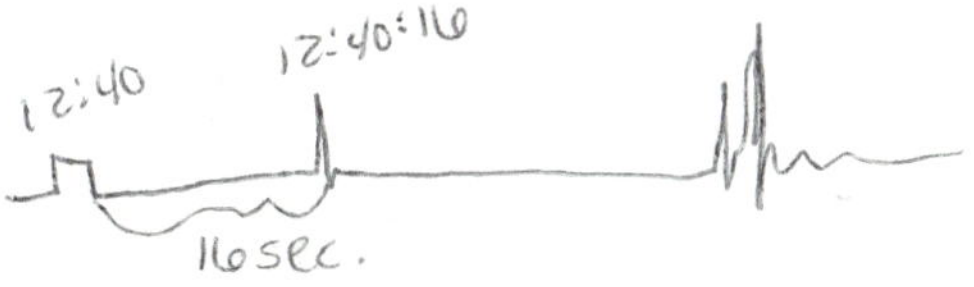

12:40 12:40:16

16 sec.

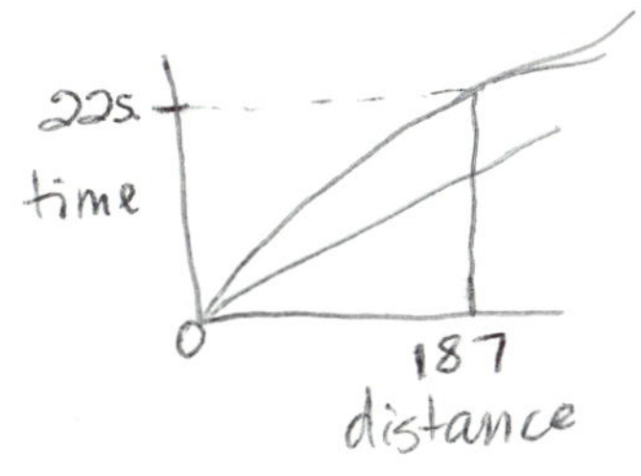

22s.

time

0 187

distance

travels 187km/22s.

12:40:16
– :22
12:39:54

163 km/20 sec. 12:40:20
– 20
12:40:00

Chapter 5
Earthquake Magnitude and Intensity*

The severity of an earthquake is generally expressed two ways: (1) by **Richter magnitude scale**, which is a measure of the energy released by the quake, and (2) by the **Modified Mercalli intensity scale**, which is subjective and describes the effects or damage by a quake at a particular location. The Richter magnitude scale was defined precisely by Charles Richter as "... the logarithm (to the base of 10) of the maximum seismic wave amplitude (in thousandths of a millimeter) recorded on a special seismograph called the Wood-Anderson, at a distance of 100 km from the earthquake epicenter." When an earthquake is recorded, the greatest deflection of the zig-zag trace on the seismogram is measured and compared to a standard reference earthquake corrected for distance to epicenter and differences in seismometers. The resulting "number" relates the size of the earthquake to the standard quake. The **standard quake** is defined as one that causes a maximum deflection of the seismograph of 0.001 mm (1 micron) at an epicentral distance of 100 km resulting in a magnitude zero quake (logarithm of 1 is zero). After the corrections for distances are made the magnitude value is constant and results in an effective means of size classification.

RICHTER MAGNITUDE SCALE

The Richter magnitude scale, named after Dr. Charles F. Richter, Professor Emeritus of the California Institute of Technology, and the scale most commonly used, is often misunderstood. On this scale, the earthquake's magnitude is expressed in whole numbers and decimals. However, magnitudes can be confusing and misleading unless the mathematical basis for the scale is understood. It is important to recognize that magnitude varies logarithmically with the amplitude of the earthquake wave recorded by the seismograph. Each whole number increase of magnitude on the scale represents an increase of 10 times the measured amplitude of the wave. Thus, the amplitude of a magnitude 8.0 earthquake is not twice as large as a shock of magnitude 4.0 but is 10,000 times larger!

---

*Source material and much of the text for this exercise came from "How Earthquakes Are Measured," California Geology, February, 1979, p. 37; Calif. Div. of Mines and Geology, Note 23.

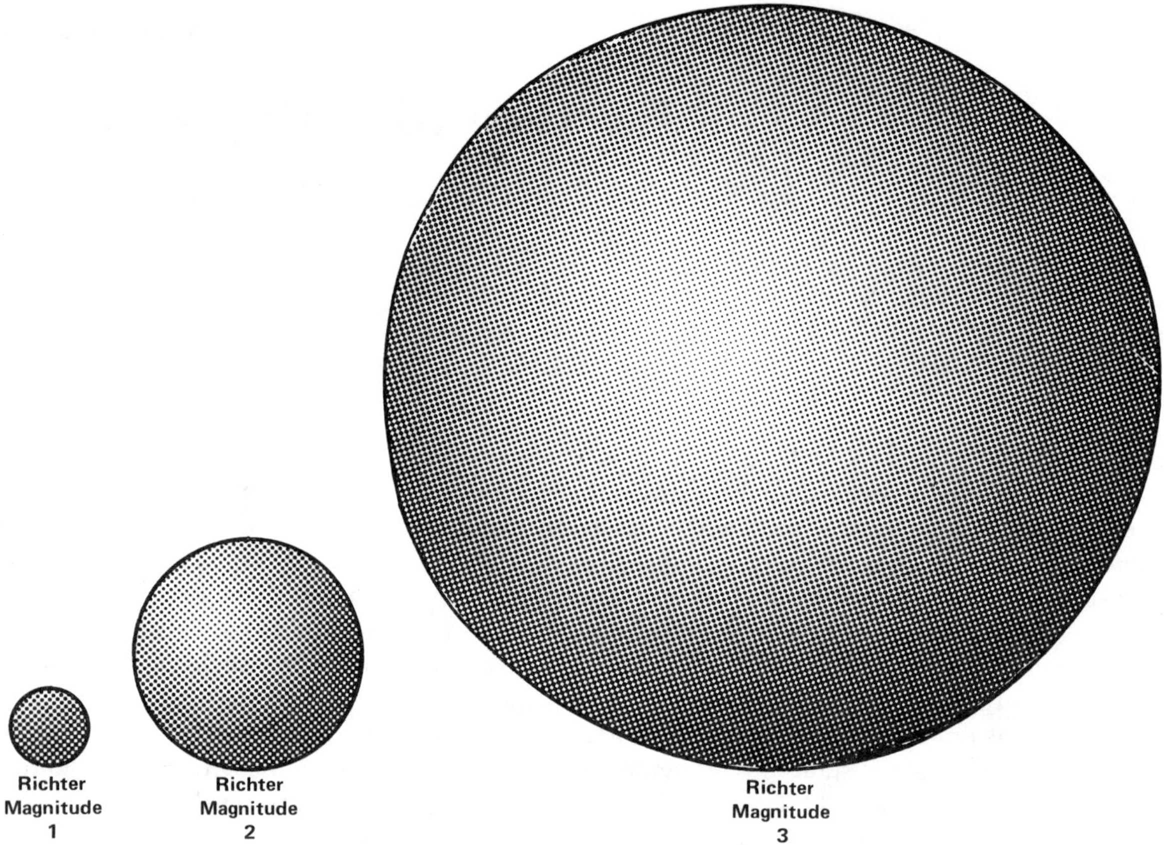

FIGURE 5-1. Relationship between earthquake magnitude and energy released. The volumes of the spheres are roughly proportional to the energy released for earthquakes of the magnitude given. The energy released by the San Francisco quake of 1906 (at the same scale) would be represented by a ball 220 feet in diameter. ("How Earthquakes Are Measured," California Geology, February, 1979, p. 35.)

Richter magnitude can also provide an estimate of the amount of **energy** released during the quake. For every unit increase in magnitude, the energy is increased by a factor of about 30. For the previous example a magnitude 8.0 earthquake releases almost one million times more energy than a magnitude 4.0 (Fig. 5-1).

A quake of magnitude 2 on the Richter scale is the smallest quake normally felt by humans. Earthquakes with a Richter magnitude of 7 or more are commonly considered to be major. The magnitude scale has no fixed maximum or minimum; observations have placed the largest recorded earthquakes in the world at about 8.9, and the smallest at –3. Earthquakes with magnitudes smaller than 2 are called "micro-earthquakes." Richter magnitudes are not used to estimate damage because an earthquake in a densely populated area, which results in many deaths and considerable damage, may have the same magnitude as one that occurs in a barren, remote area, which may do nothing more than frighten the wildlife.

MODIFIED MERCALLI INTENSITY SCALE

The first scale to reflect earthquake intensities was developed by de Rossi of Italy, and Forel of Switzerland, in the 1880s. This scale, with values from I to X, was used for about two decades. A need for a more refined scale increased with the advancement of the science of seismology, and in 1902 the Italian seismologist Mercalli devised a new scale on a I to XII range. The Mercalli scale was modified in 1931 by American Seismologists Harry O. Wood and Frank Neumann to take into account modern structural features.

I. Not felt except by a very few under especially favorable circumstances.

II. Felt only by a few persons at rest, especially on upper floors of buildings. Delicately suspended objects may swing.

III. Felt quite noticeably indoors, especially on upper floors of buildings, but many people do not recognize it as an earthquake. Standing motor cars may rock slightly. Vibration like passing of truck. Duration estimated.

IV. During thc day felt indoors by many, outdoors by few. At night some awakened. Dishes, windows, doors disturbed, walls make cracking sound. Sensation like heavy truck striking building. Standing motor cars rocked noticeably.

V. Felt by nearly everyone, many awakened. Some dishes, windows, etc. broken; a few instances of cracked plaster; unstable objects overturned. Disturbances of trees, poles and other tall objects sometimes noticed. Pendulum clocks may stop.

VI. Felt by all, many frightened and run outdoors. Some heavy furniture moved; a few instances of fallen plaster or damaged chimneys. Damage slight.

VII. Everybody runs outdoors. Damage negligible in buildings of good design and construction; slight to moderate in well-built ordinary structures; considerable in poorly built to badly designed structures; some chimneys broken. Noticed by persons driving motor cars.

VIII. Damage slight in specially designed structures; considerable in ordinary, substantial buildings, with partial collapse; great in poorly built structures. Panel walls thrown out of frame structures. Fall of chimneys, factory stacks, columns, monuments, walls. Heavy furniture overturned. Sand and mud ejected in small amounts. Changes in well water. Persons driving motor cars disturbed.

IX. Damage considerable in specially designed structures; well-designed frame structures thrown out of plumb; great in substantial buildings, with partial collapse. Building shifted off foundations. Ground cracked conspicuously. Underground pipes broken.

X. Some well-built wooden structures destroyed; most masonry and frame structures destroyed with foundations; ground badly cracked. Rails bent. Landslides considerable from river banks and steep slopes. Shifted sand and mud. Water splashed (slopped) over banks.

XI. Few, if any, (masonry) structures remain standing. Bridges destroyed. Broad fissures in ground. Underground pipelines completely out of service. Earth slumps and land slips in soft ground. Rails bent greatly.

XII. Damage total. Practically all works of construction are damaged greatly or destroyed. Earthquake waves seen on ground surface. Lines of sight and level are distorted. Objects are thrown upward into the air.

The Modified Mercalli intensity scale measures the intensity of an earthquake's effects in a given locality, and is perhaps much more meaningful to the layman because it is based on actual observations of earthquake effects at specific places. It should be noted that because the data used for assigning intensities can be obtained only from direct, firsthand reports, considerable time—weeks or months—is sometimes needed before an intensity map can be assembled for a particular earthquake. On the Modified Mercalli intensity scale, values range from I to XII. The most commonly used adaptation covers the range of intensity from the conditions of "I—not felt except by very few, favorably situated," to "XII—damage total, lines of sight disturbed, objects thrown into the air." Whereas an earthquake has only one magnitude, it can have many intensities which decrease with distance from the epicenter.

COMPARISON OF MAGNITUDE AND INTENSITY

It is difficult to compare magnitude and intensity because intensity is linked with the particular ground and structural conditions in a given area, as well as distance from the epicenter; whereas magnitude depends on the energy released at the focus of the earthquake. For example, in the San Francisco earthquake of 1906 a house directly on the San Andreas fault had its front porch sheared off but was otherwise little damaged, while buildings many miles away in the city constructed on bay muds were totally destroyed. Table 5-1 gives an approximate comparison of magnitude and intensity.

TABLE 5-1
RICHTER MAGNITUDE AND MODIFIED MERCALLI INTENSITY COMPARED*

Richter Magnitude	Modified Mercalli Intensity	Description
2	I–II	Usually detected only by instruments
3	III	Felt indoors
4	IV–V	Felt by most people; slight damage
5	VI–VII	Felt by all, many frightened and run outdoors; damage minor to moderate
6	VII–VIII	Everybody runs outdoors, damage moderate to major
7	IX–X	Major damage
8+	X–XII	Total and major damage

*After Charles F. Richter, 1958, Elementary Seismology, San Francisco: W. H. Freeman and Co.

EXERCISE V

EARTHQUAKE MAGNITUDE AND INTENSITY

Name ______________________

1. A **nomogram** is a graph that enables you by the aid of a straightedge to read off the value of a dependent variable when the values of two or more independent variables are given. Using the nomogram of Figure 5-2, determine the Richter magnitude of the following earthquakes:

	S - P	Amplitude	Magnitude
(a)	2 sec	20 mm	2.9
(b)	10 sec	20 mm	4.3
(c)	50 sec	20 mm	6
(d)	20 sec	0.2 mm	2.7
(e)	20 sec	2.0 mm	3.7
(f)	20 sec	20.0 mm	4.7

2. For problems (a–c) of Question 1, the amplitude of the seismic wave remains constant but the distance to epicenter (as measured by the difference in travel time between S-waves and P-waves (S – P)) changes. Can you identify and state a simple relationship between distance to epicenter (S – P), amplitude, and magnitude for these problems?

the greater the distance to epicenter-
the greater the magnitude.

3. For problems (d – f) of Question 1 the value of S – P remains constant, but the amplitude changes by a factor of 10. Can you identify a simple relation between S – P, amplitude, and magnitude for these problems?

when amplitude is increased by ten
the magnitude will increase by one.

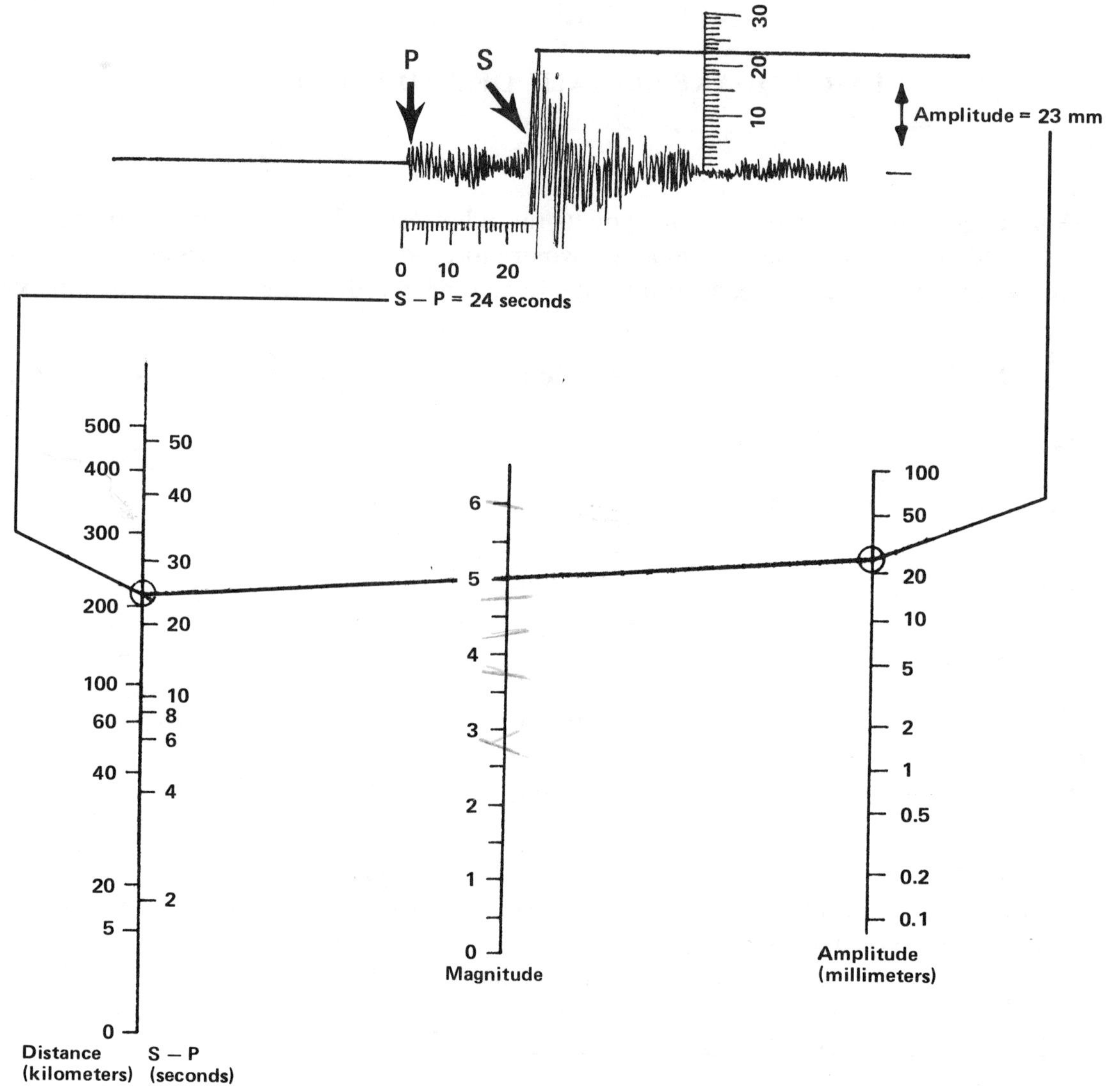

PROCEDURE FOR CALCULATING THE LOCAL MAGNITUDE, M_L

1. Measure the distance to the focus using the time interval between the S and the P waves (S – P = 24 seconds).

2. Measure the height of the maximum wave motion on the seismogram (23 millimeters).

3. Place a straightedge between appropriate points on the distance scale (left) and amplitude scale (right) to obtain magnitude M_L = 5.0.

FIGURE 5-2. Nomogram for determining Richter magnitude (M_L) (from Bolt, Bruce, 1977, (from Earthquakes: A Primer, by Bruce A. Bolt. W. H. Freeman and Co. Copyright © 1978).

The objectives of the following questions are to become familiar with the Modified Mercalli scale and to construct an isoseismal map. **Isoseismals** are lines that connect points of equal earthquake intensity and enclose areas of approximately equal earthquake damage. This discloses something about location of the epicenter and local ground conditions. In order to gather information on intensities, questionnaires are sent out requesting information on the effects of an earthquake in a large area surrounding the epicenter (Fig. 5-3).

4. The following descriptions are personal reports on effects of a disastrous earthquake of February 9, 1971, at San Fernando, California. Use these descriptions to determine the local intensities (Table 5-1) and plot these intensities on the map (Fig. 5-4) at the given locations. Then construct isoseismal lines dividing the map into compartments according to intensity. The lines need not close and can just be drawn off the limits of the map. For ease of construction, combine intensity units as follows: 0–III, IV–V, VI–VII, VIII, IX, X–XII.

North on I-5 from downtown Los Angeles

Los Angeles (downtown)	— old Midnight Mission suffered major damage; major damage also at L.A. High School.
Burbank	— mayor said only old, poorly designed buildings suffered much damage.
Sunland	— some ground cracks; some buildings off foundations; many chimneys fallen.
San Fernando	— old part of Veterans' Hospital collapsed; new Olive View Hosp. damaged beyond repair; railroad tracks bent and broken.
Newhall	— ground cracks; partial collapse of a few buildings.
Castaic	— considerable damage to a few old buildings.
Gorman	— some cracked plaster. People glad it hadn't been San Andreas fault.

Take Highway 99 north

Bakersfield	— noticed cracked plaster on old Bakersfield Inn; quake was felt by most everyone who was up.
Pixley	— everyone who was awake felt quake, but damage was just some cracked plaster.
Tulare	— some people awakened by quake; no damage.
Kingsburg	— only a few people on upper floors felt the quake.
Caruthers	— Dave Harder was up early. He felt the quake, saw his car move, and heard the house creaking.
Fresno	— people on upper floors of tall buildings felt quake; hanging lamps swayed.

Return south on I-5

Coalinga	— felt by a few people indoors.
Buttonwillow	— many people awakened; some cracked plaster.

Northwest on U.S. 101 toward San Francisco

Hollywood	— saw considerable damage to old buildings and one broken chimney.

(Data continued on p. 62)

NOS FORM 680 (11-70)

U.S. DEPARTMENT OF COMMERCE
NATIONAL OCEANIC AND ATMOSPHERIC ADMINISTRATION
NATIONAL OCEAN SURVEY

OMB No. 41-R0013
Approval Expires June 30, 1975

EARTHQUAKE REPORT

1. An earthquake was felt [X]; not felt [ ]

Date of shock February 9, 1971

Time 0601 A.M.

_____ P.M.

If felt, please supply information below *(Underline appropriate words or fill spaces.)*
If not felt, please sign and return card, which requires no postage.

2. YOUR LOCATION DURING EARTHQUAKE

a. City, County, State, and Township L.A. County, Calif.

Range, Section, Quarter Section, or Coordinates R-3N-14W. Sec. 32

b. Ground:

Rocky, gravelly, loose, compact, marshy, filled in, or _____

Level, sloping, steep, or _____

c. If inside, type of construction

Wood, brick, stone, or wood frame

d. Quality of construction

New, old, well built, poorly built, or _____

e. No. of floors in building: Single

f. Observer's floor: 1st

g. Activity when earthquake occurred: Walking, sitting, lying down, sleeping

h. If outside, you, others were: Quiet, active

3. EFFECTS ON POPULATION

a. Felt by:

Very few, several, many, all (in your home) (in community)

b. Awakened:

No one, few, many, all (in your home) (in community)

c. Frightened:

No one, few, many, all (in your home) (in community) general panic

4. RELATED SOUNDS

a. Rattling of windows, doors, dishes, etc. All of above prior to shock

b. Creaking of building (Describe) Loud-snapping

c. Earth noises: Faint, moderate, loud Loud-similar to Jet

5. PHYSICAL EFFECTS AND DAMAGE

a. Outside:

(1) Trees and bushes shaken, vehicles rocked, etc. Yes

(2) Ground cracked; landslides; water disturbed, etc. All of these

(3) Chimneys, tombstones, elevated water tanks, etc., cracked, twisted, overturned All of these

(4) Other effects _____

b. Buildings:

(1) Hanging objects swung moderately, violently Direction East to West

(2) Small objects shifted, overturned, fell Yes

(3) Furniture shifted, overturned, broken Yes

(4) Plaster cracked, broken, fell Yes

(5) Windows cracked Yes

(6) Structural elements of brick, wood, or Wood

Damage slight, moderate, great Great

Signature and address of observer

Scott E. Franklin Eng. 74
Capt. LA County Fire Dept.

12587 N. Dexter Pk Rd
San Fernando
(Kagel Cyn.)

Additional information will be appreciated. Use space on reverse side.

FIGURE 5-3. Questionnaire for determining earthquake intensity indicates violent shaking in Kagel Canyon, February 9, 1971.

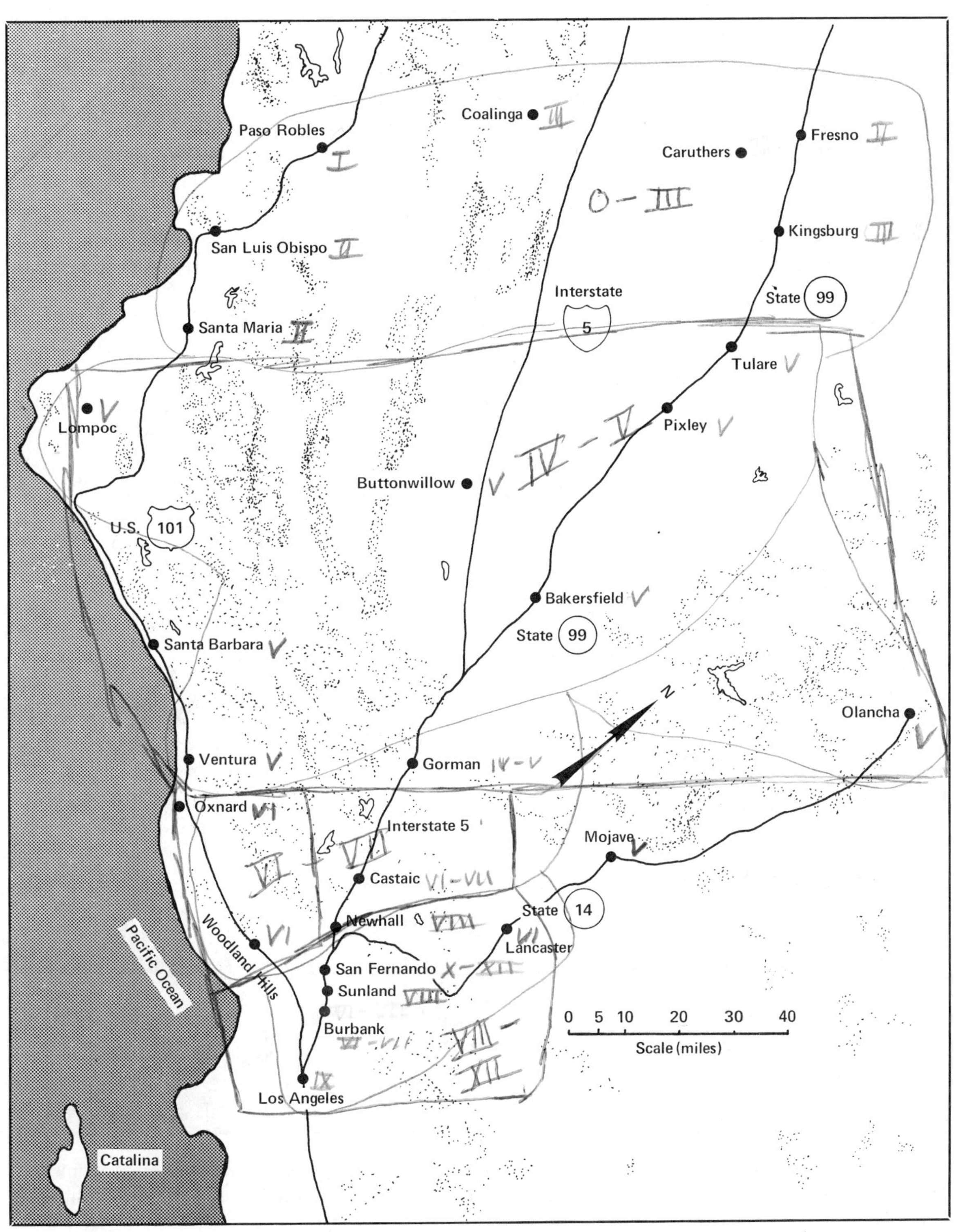

FIGURE 5-4. Base map for plotting isoseismals.

(Continued from p. 59)

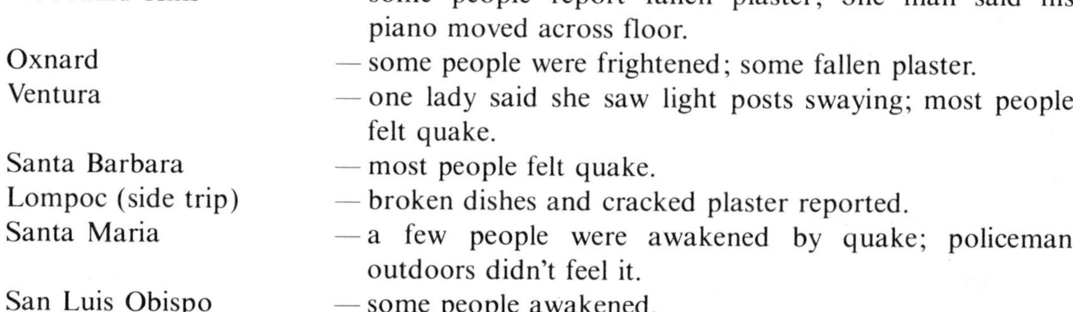

Woodland Hills	— some people report fallen plaster; one man said his piano moved across floor.
Oxnard	— some people were frightened; some fallen plaster.
Ventura	— one lady said she saw light posts swaying; most people felt quake.
Santa Barbara	— most people felt quake.
Lompoc (side trip)	— broken dishes and cracked plaster reported.
Santa Maria	— a few people were awakened by quake; policeman outdoors didn't feel it.
San Luis Obispo	— some people awakened.
Paso Robles	— not felt.

North on Hwy 14 from Palmdale

Palmdale	— everyone felt quake; some heavy furniture moved; fallen plaster reported.
Lancaster	— everyone felt quake; fallen plaster.
Mojave	— many people awakened; a few broken windows.
Olancha	— cracked plaster; most people awakened.

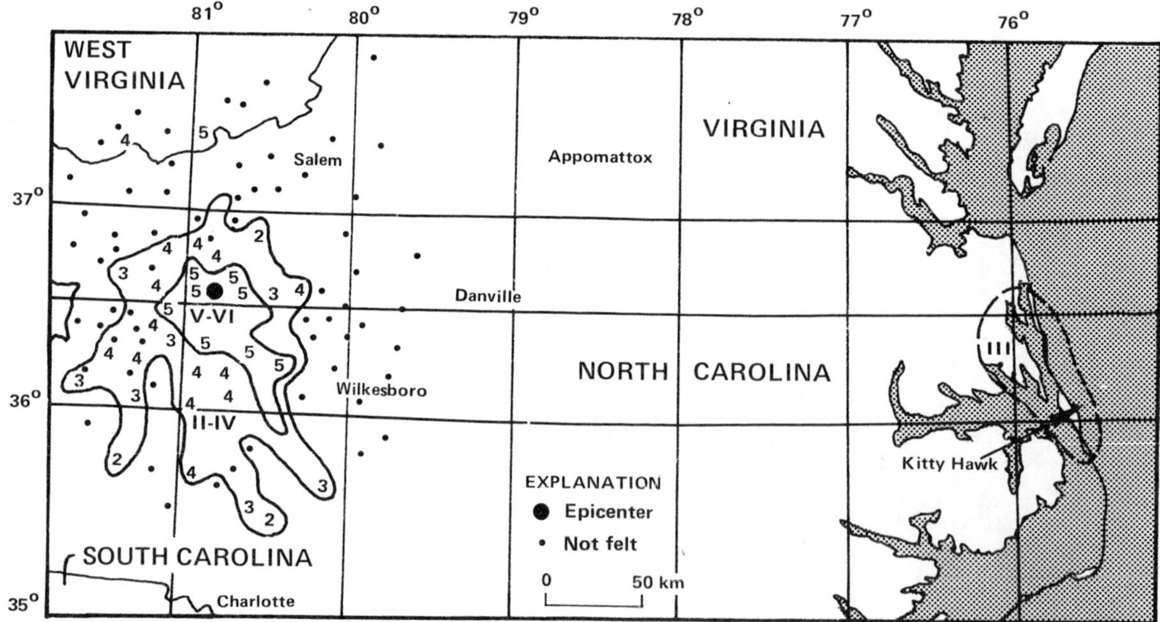

FIGURE 5-5. Isoseismal map for the Virginia earthquake of 13 September 1976. Roman numerals represent Modified Mercalli intensities between isoseismals. Arabic numerals represent these intensities at specific sites.

5. Figure 5-5 shows an isoseismal map for the earthquake of September 13, 1976, in Virginia. Roman numerals represent the Modified Mercalli intensities.

   (a) Give a reasonable explanation for the irregular pattern of isoseismals around the epicenter.

   the intensity of a quake often depends on geology at site

   (b) Explain the isoseismal line in the area of Kitty Hawk, North Carolina, about 450 km from the epicenter, and the absence of any effects of the quake between the 2 areas.

   maybe there was no one there to feel it.
   water in sand- liquification

Chapter 6

Seismic Hazards and Land-Use Planning

Urban planners, public officials, and all people who live in seismic regions are concerned about minimizing loss of life and property damage due to earthquakes. In California alone, it is estimated that there will be a 20-billion-dollar property loss due to earthquakes by the year 2000 if no precautionary measures are taken. On the other hand, if all possible measures to reduce earthquake losses are undertaken, then property damage will be half the estimated amount. The cost of such mitigating measures would be about 2 billion dollars but would result in a savings of 10 billion dollars. A semi-quantitative measure of this kind of government action is called the **benefit-cost ratio**, which in the case cited is 5—a very attractive course of action indeed.

In order to refine estimates of earthquake damage and potential hazards, the planner and official must have certain types of geologic information available, mostly in the form of maps. Maps showing the locations of active faults and historic fault rupture, geologic materials, slopes and landslides, and depth to water table are but a few examples.

SEISMIC HAZARDS

Earthquakes give rise to various geologic phenomena that cause severe property damage and loss of life. There are five seismic-related processes that should be considered in any preliminary site investigation by geologists and engineers in earthquake country. The detail and thoroughness of the study depends upon the project. Obviously critical facilities such as nuclear power plants, liquefied natural gas facilities, and power generating stations require extensive and intensive site studies. The five processes generated by earthquakes are: (1) surface fault rupture, (2) ground shaking, (3) landslides, (4) foundation (ground) failures not related to landslides, and (5) tsunami and seiche near the coast of large bodies of water. We will consider these in some detail; tsunami will be discussed again in Chapter 11.

SURFACE RUPTURE

Some active faults, such as the San Andreas fault in California, exhibit surface faulting. There is no way to design a building against such displacement should a structure be built directly on a fault. We therefore must learn to recognize the presence of active faults at the ground surface and restrict construction in those areas. The definition of an active fault varies according to the type of land use contemplated or the importance of the structure. For example, the State of California defines an active fault as one that has moved within the last 11,000 years (within Holocene time). The Nuclear Regulatory Commission, on the other hand, considers a fault active if it has moved within the last 35,000 years, or more than once within the past 500,000 years. The latter definition is more restrictive and is in keeping with the greater hazard potential of facilities like atomic reactors in contrast to commercial or residential development projects.*

Types of Faults. Faults may be defined on the basis of the geometry of the fault surface and the relative motion of rock masses on either side of the fault. Figure 6-1 shows that for a non-vertical fault the hanging wall (upper block) may move relative to the footwall (lower block) up or down the dip (inclination from the horizontal) of the fault. Such movement is known as **dip slip**. Also note from the figure that movement may take place parallel to the trend or strike of the fault; such movement is called **strike slip**. Thus we may recognize three types of faults and fault movement as follows:

(1) **Reverse slip faults**, in which the hanging wall has moved up relative to the footwall. Low-angle reverse faults are called **thrust faults** and occur at sites of intense compression in the Earth's crust, such as subduction zones (Fig. 6-2a).

(2) **Normal slip faults**, in which the hanging wall has moved down relative to the footwall. Normal faults are the result of extension or pulling apart of the Earth's crust (Fig. 6-2b).

(3) **Strike-slip faults**, where the dominant movement has been parallel to the strike or trend of the fault. Faults of this type may mark active or ancient plate boundaries where the plates are slipping past one another. The relative motion on these faults may be **sinistral** (left-slip) (Fig. 6-2c) or **dextral** (right-slip) (Fig. 6-2d).

Fault Recognition. Recognition of active faults may be based upon the historical record, sesimic activity, landforms indicative of recent faulting such as scarps, sag ponds, offset

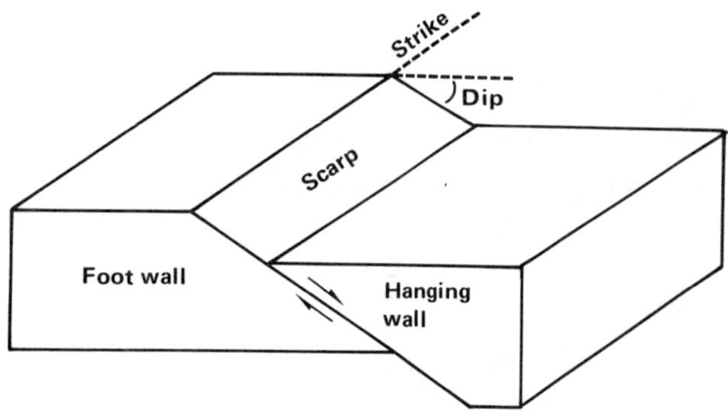

FIGURE 6-1. Block diagram illustrating fault nomenclature and geometry.

*For a good discussion of land use in seismic regions, see Nichols, D.R., and Buchanan-Banks, J.M., 1974, Seismic Hazards and Land-Use Planning; U.S. Geological Survey Circular 690.

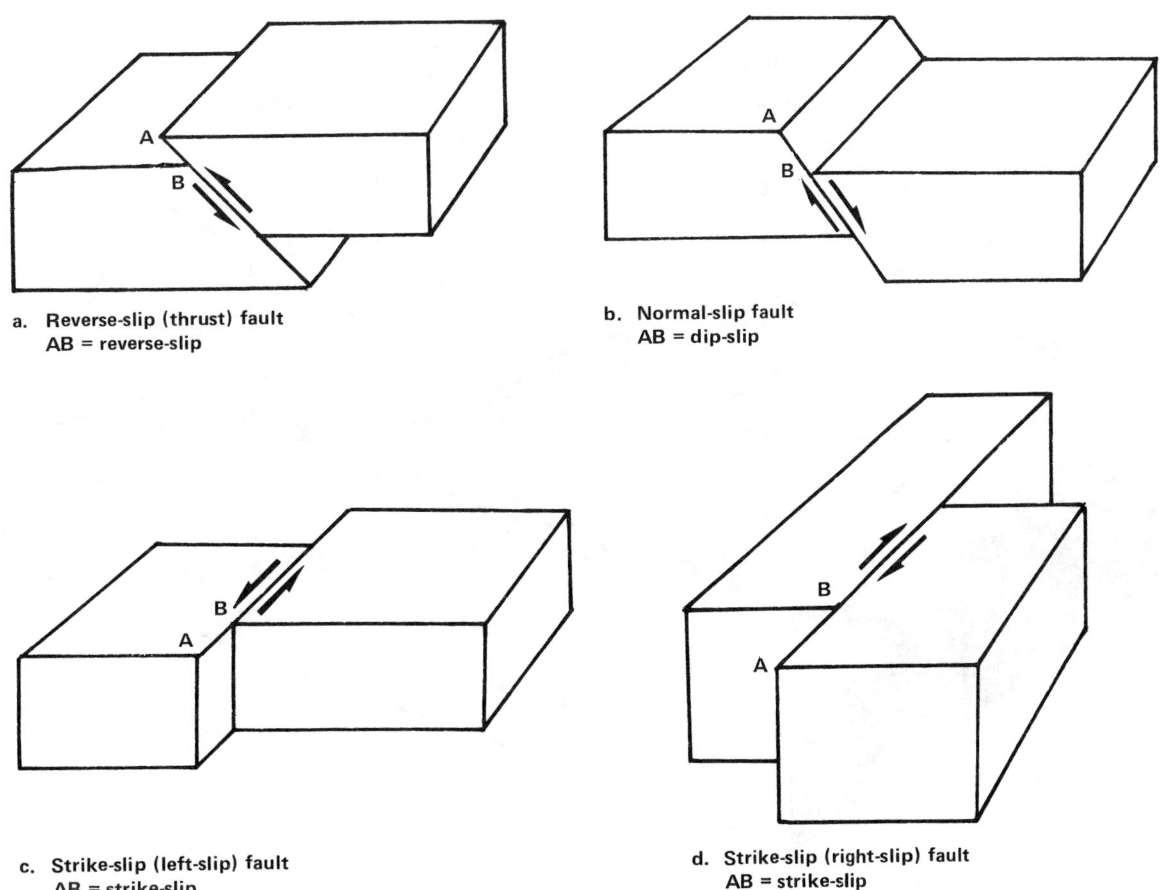

FIGURE 6-2. Relative motion on three common types of faults.

stream courses, or disruption of man-made features such as fence lines and curbs. Figure 6-3 illustrates the common recognizable surface expression of active or recently active faults. It should be noted that most major faults are made up of hundreds of fault surfaces in a "zone" varying in width from a few hundred to many thousands of feet. Thus, geologists charged with protecting the public must identify such hazardous areas and restrict development in them (Fig. 6-4).

Rupture Hazard Assessment. In order to assess rupture hazard it is necessary to know the exact location of a particular fault or fault zone. Many active faults in California have been mapped in detail on large-scale maps and are used in planning. A portion of such a map is shown in Figure 6-5. Where a fault consists of a wide zone of fractures, geologists generally assume that the surface on which the fault last slipped is where it will slip again. The next step is to estimate how big an earthquake might be generated on that fault, called the **maximum credible earthquake**, and then attempt to evaluate how often such an event is likely to occur. Data have been compiled relating total rupture length on a fault and maximum displacement of a single event to earthquake magnitude (Figs. 6-6 and 6-7). The practice has been to determine the total fault length in the field and then assume that one-half the total length would be activated during a maximum credible earthquake on that fault. Maximum dis-

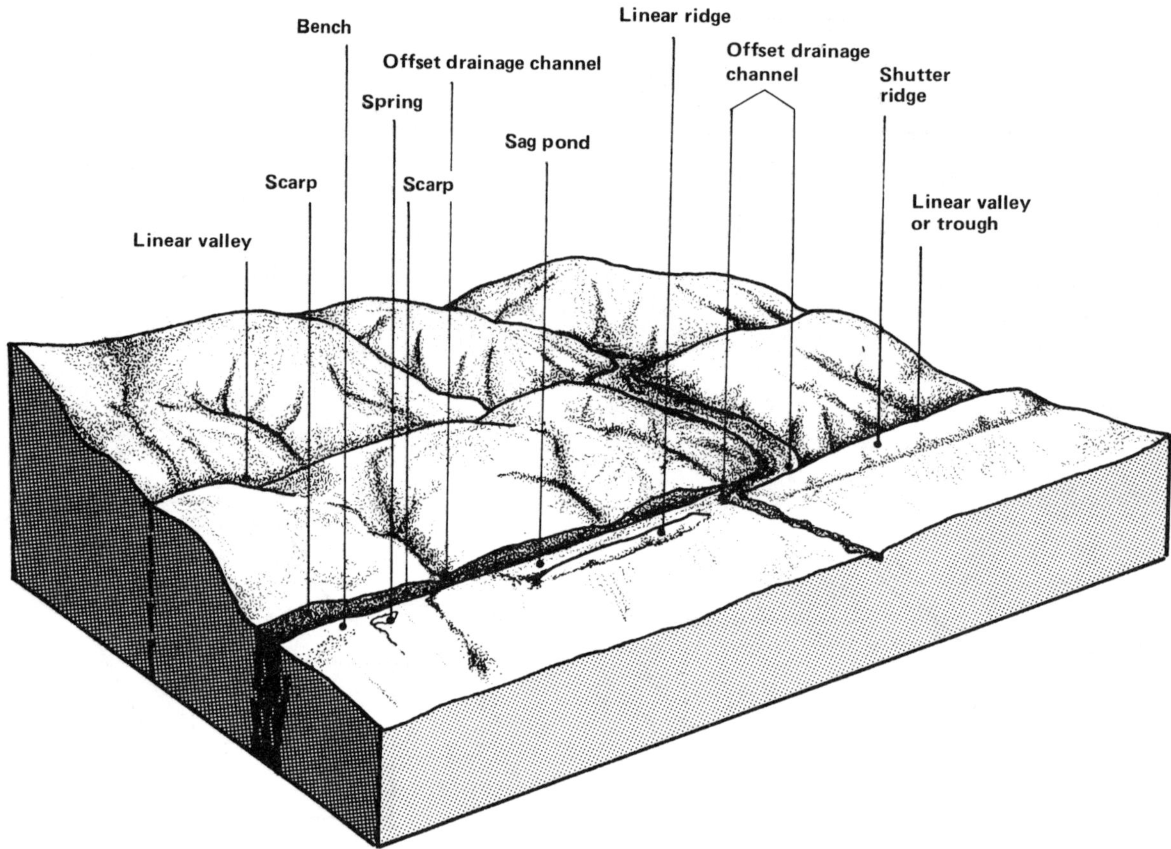

FIGURE 6-3. Block diagram showing landforms produced along recently active faults (from Vedder, J. G., and Wallace, R.E., 1970, U.S. Geol. Survey, Misc. Geologic Investigations Map I-574).

placement values can be used if the historic record is long enough or if trenching or exposures indicate what is likely to be the maximum displacement (offset) on the fault in question. Thus from Figure 6-6 we can see that the maximum credible earthquake on a fault 100 miles long (50-mile rupture length) would be 7.4, but might range from magnitude 6.9 to 7.9.

Recurrence interval or frequency is quite another problem. One statistical approach is shown in Figure 6-8 for southern California. The frequency of a magnitude 7 earthquake, for instance, is about 0.1 earthquake/year. This means we can expect such a quake once every ten years or there is a 10 percent chance of such an event occuring in any one year. This plot describes the seismicity of a region and has been assembled for the entire world as well as for smaller areas where there is sufficient historical data. This type of information has limited use for planning because it only tells us the statistical chance of a certain magnitude earthquake occurring within the study area.

The second method of determining recurrence intervals of earthquake events is based upon estimating the average annual strain buildup in rocks adjacent to the fault from measurements across the fault. We may then use the maximum known displacement along the fault to determine both the maximum credible quake and the recurrence interval. The logic is quite simple. If the rocks on opposite sides of a fault are being bent at a rate of 2 cm/year and the maximum displacement ever to have occurred on that fault is thought to be 200 cm (about 6 ft), it would take about 100 years for this amount of strain to build up, at which time the rocks would break. Thus,

$$\text{Recurrence interval (R)} = \frac{\text{Maximum expectable displacement (d)}}{\text{Long-term strain build-up (s)} - \text{Creep (c)}}$$

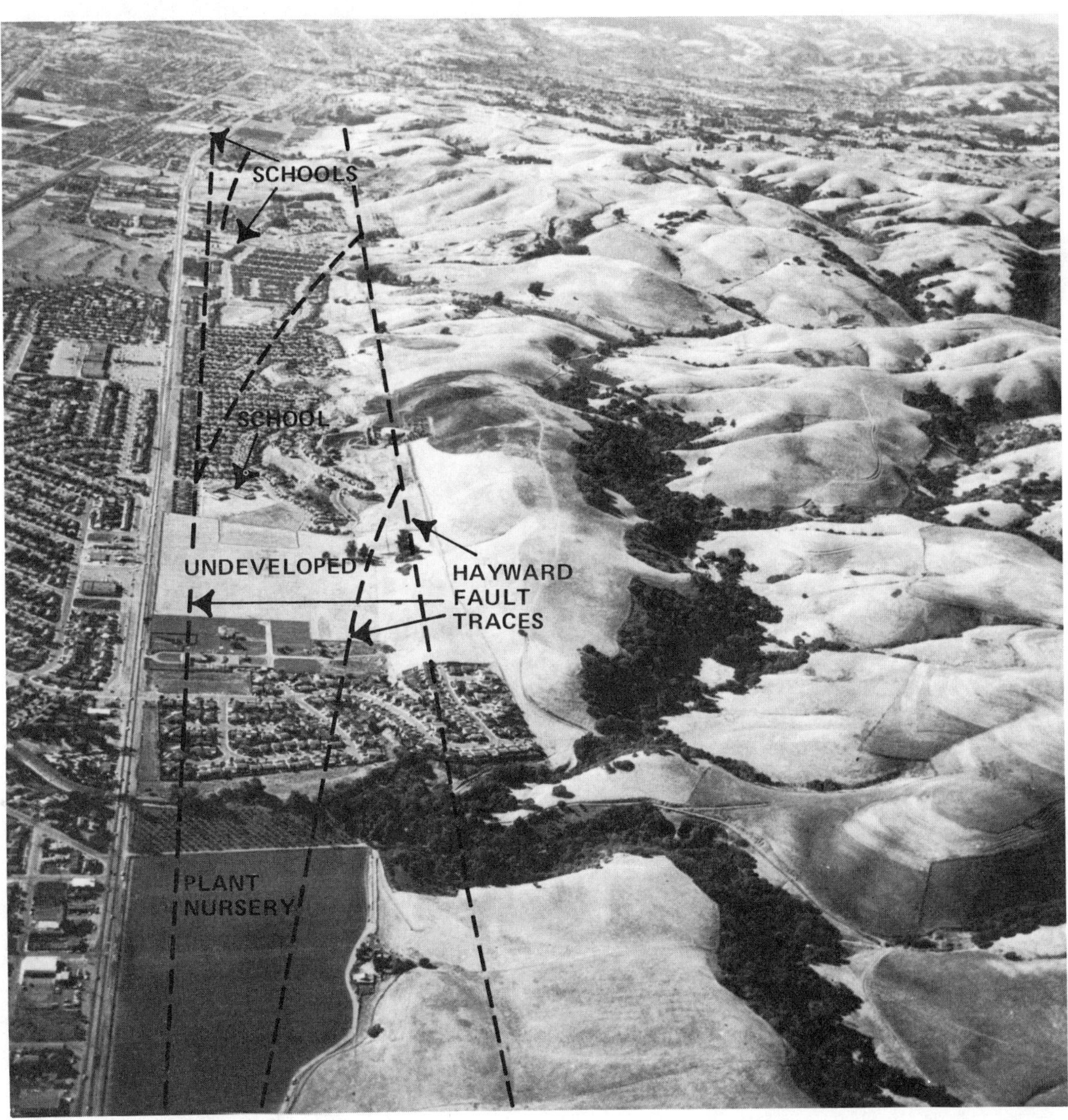

FIGURE 6-4. Air photo showing approximate location of fault traces within the Hayward fault zone. Some of these ruptured in the October 21, 1868, earthquake. (Nichols, D.R., and Buchanan-Banks, J.M., 1974, Seismic Hazards and Land-Use Planning: U.S. Geol. Survey Circular 690.)

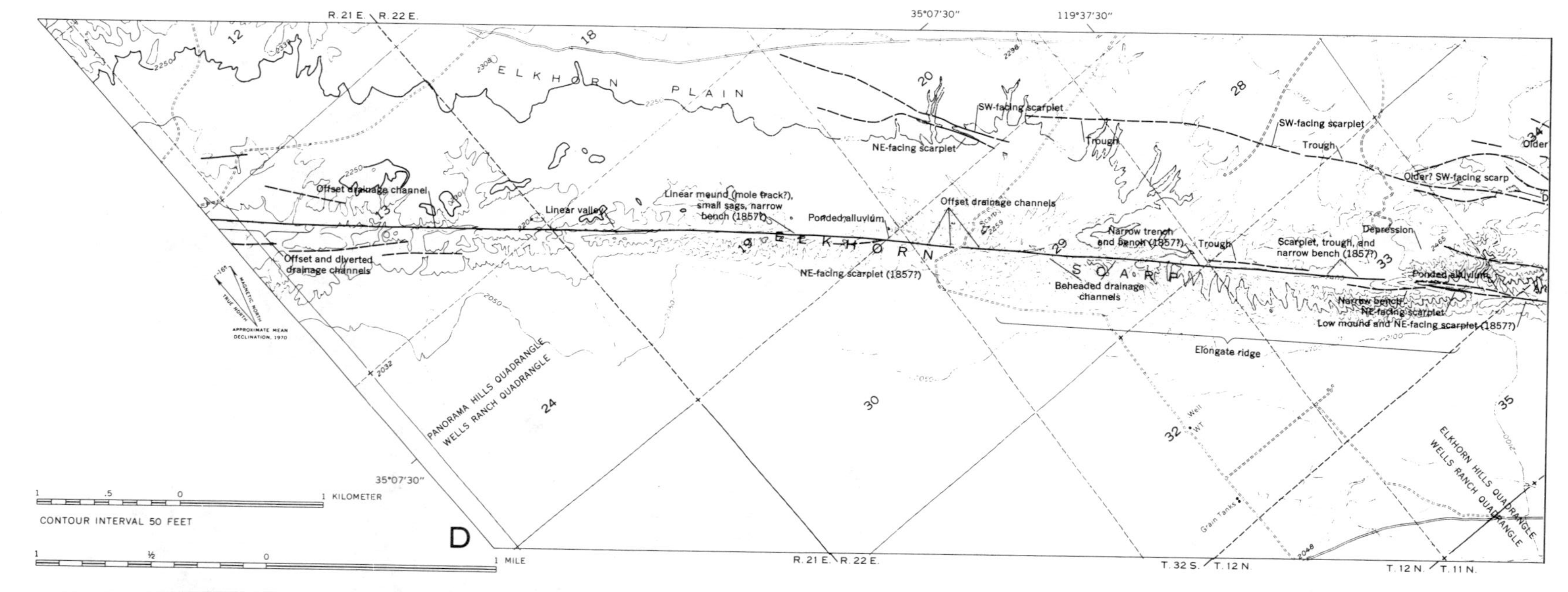

FIGURE 6-5. Map showing recently active breaks along the San Andreas fault in the vicinity of Elkhorn Hills, San Luis Obispo County, California. (After Vedder, J.G., and Wallace, R.E., 1970, U.S. Geol. Survey, Misc. Geologic Investigations Map I-574.) The inset photos are of sections of the fault in the same area as the map. Note particularly the Elkhorn scarp and the narrow trench marking the fault.

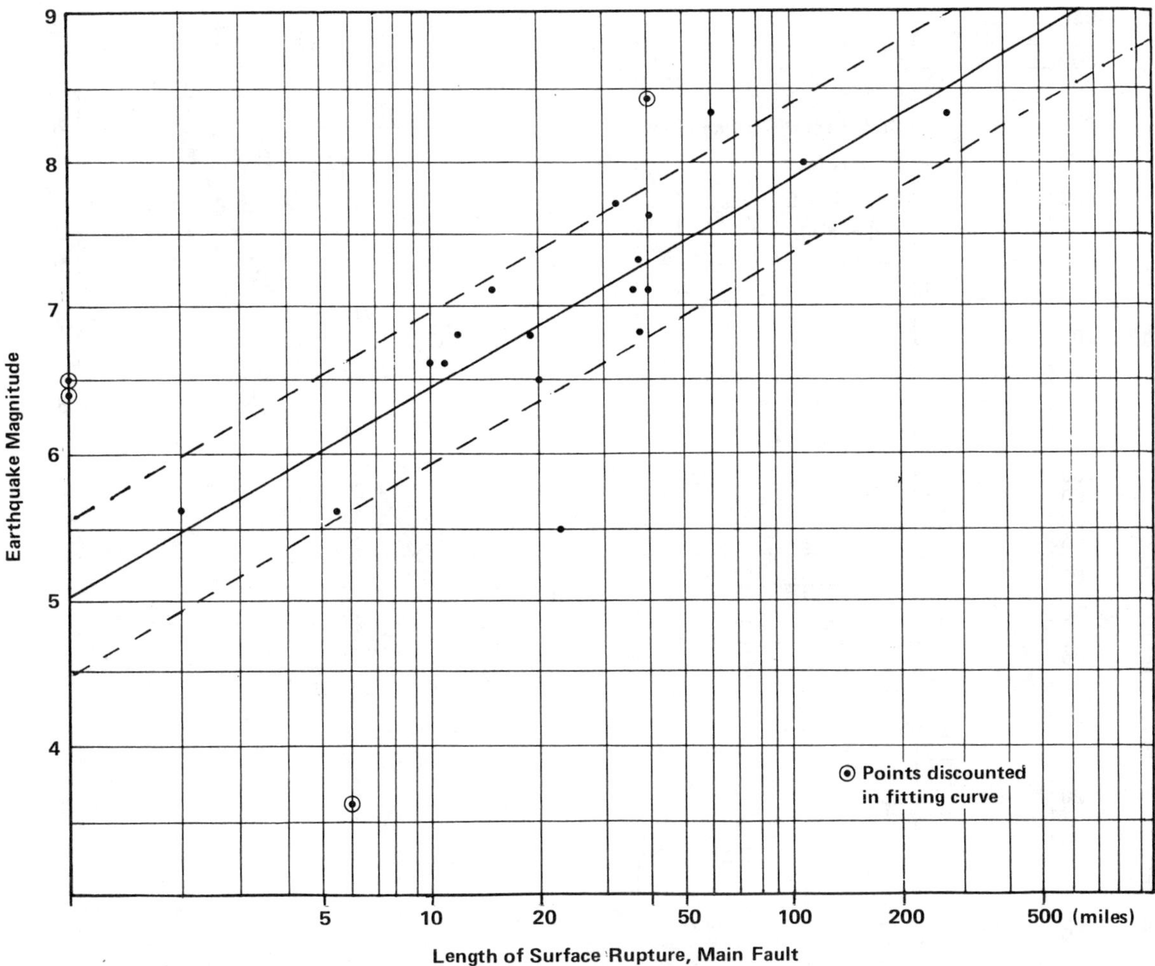

FIGURE 6-6. Earthquake magnitude versus fault rupture length. Dashed lines show the probable range of values. (Modified from Greensfelder, R.A., 1974, Maximum Credible Rock Acceleration from Earthquakes in California, California Division Mines and Geology, Map Sheet 23.)

Fault creep is slow movement on a fault without attendant strain accumulation (or quakes). This occurs on many faults in the San Andreas system at a rate of millimeters per year resulting in disrupted buildings, curbs and sidewalks where the fault trends through urban areas (Fig. 6-9). Because creep reduces strain accumulation it has to be subtracted in the denominator of the recurrence interval equation.

As an example of a practical application of the method let us assume we are interested in a fault with a strain rate of 2 cm/year and a maximum credible earthquake of magnitude 7 based upon rupture of one-half the entire fault length (Fig. 6-6). From Figure 6-7 we see that

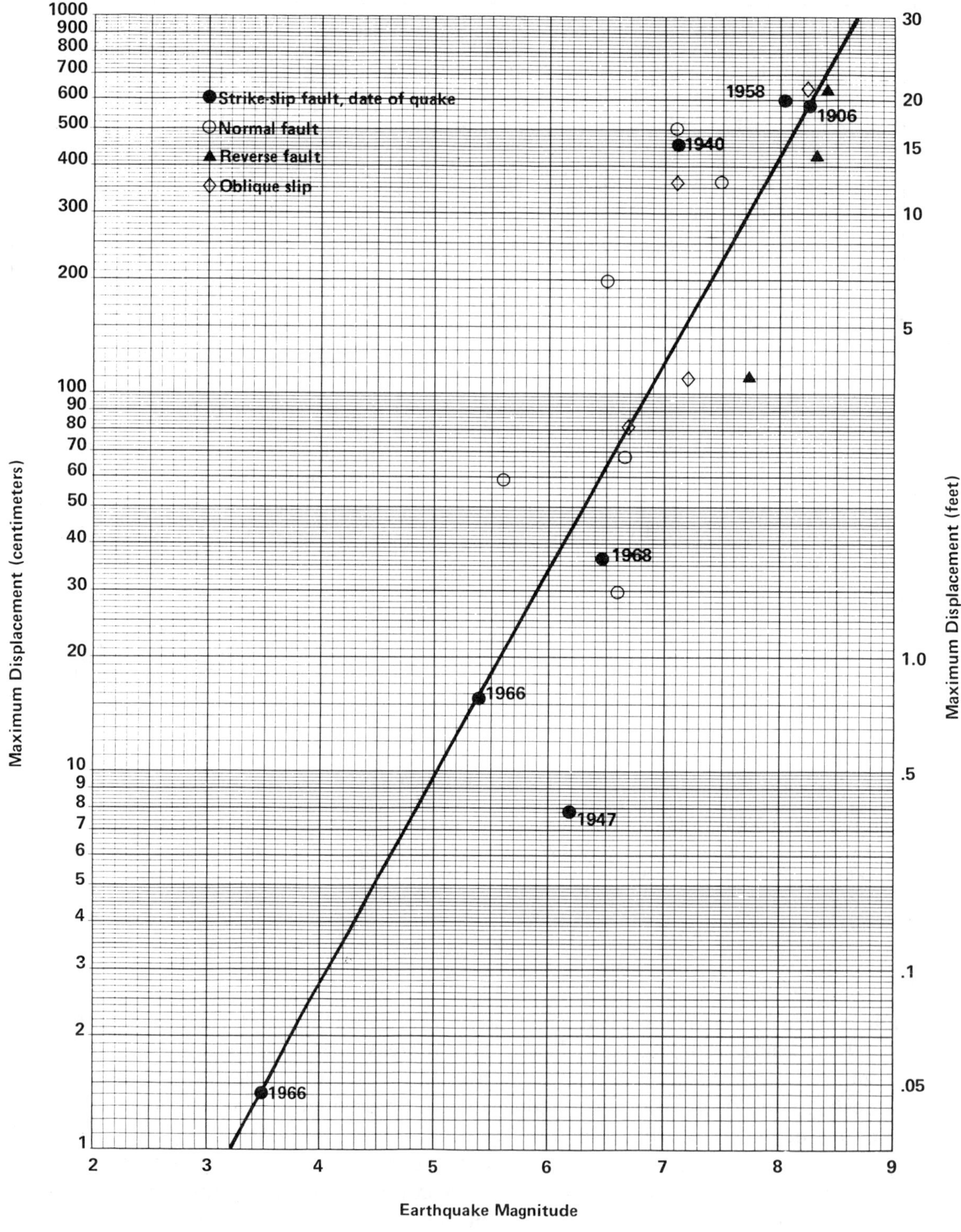

FIGURE 6-7. Magnitude-displacement (offset) relationship in historic faulting in North America.

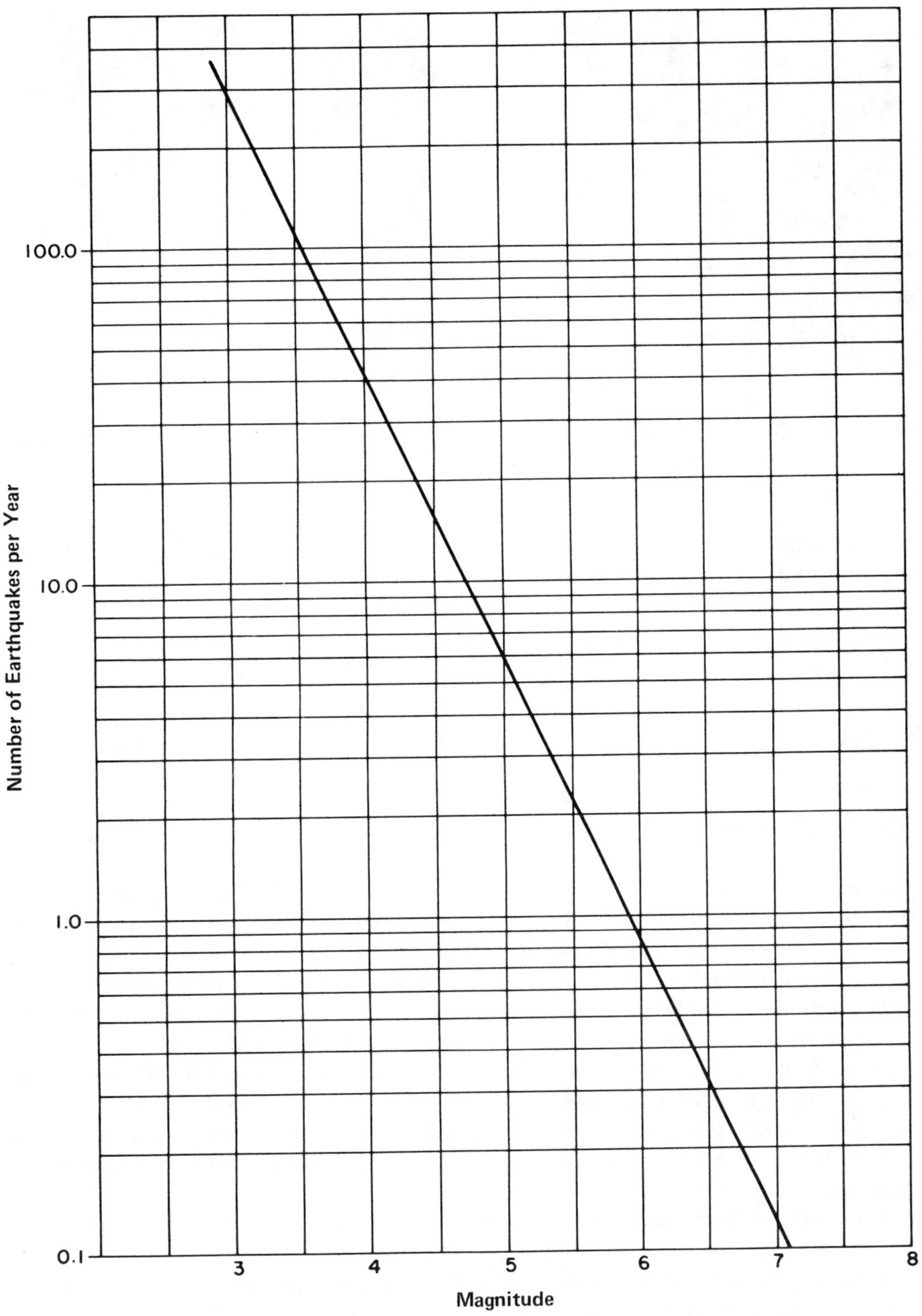

FIGURE 6-8. Earthquake magnitude versus frequency of quake for southern California.

FIGURE 6-9. Creep effects on the Calaveras fault (part of the San Andreas fault system) in Hollister, California.

a quake of magnitude 7 would produce a displacement of 1 m. Now the recurrence interval for the fault in question may be estimated (assuming no creep) as

$$R = \frac{100 \text{ cm}}{2 \text{ cm/year}} = 50 \text{ years}$$

Although the method is crude it does offer some promise for planning purposes.

GROUND SHAKING

Ground shaking is one of the most difficult seismic hazards to predict. In general, coherent solids (bedrock) shake less than alluvium or water-saturated soft sediments. Other factors include distance to epicenter, kind of fault motion, and earthquake magnitude. A first approach to this problem would be to map earth-material firmness and outline high and low hazard areas. Intensity maps are useful because the distribution of isoseismals (see Exercise V) may disclose areas of soil or sediment that would be poor foundation materials (Fig. 6-10). Where intensity maps do not exist for previous earthquakes or data are limited, they may be prepared using a method developed recently in the Soviet Union. Increments of intensity (ground shaking) are added to a base intensity computed for a design earthquake. The value of the intensity increments depends on the local ground conditions—soft ground receives higher intensity increments than firm ground.

Another factor in shaking damage is horizontal acceleration or "base shear." High accelerations are analogous to having the rug you are standing on pulled away very rapidly. Buildings are constructed to withstand vertical loads, but must be specially designed to take rapid horizontal ground motion. Many earthquakes of moderate magnitude have generated very high accelerations and thus resulted in extensive property damage. The measure of this motion is the acceleration of gravity (980cm/sec^2) and for earthquakes it is expressed as a **percentage of gravity**

FIGURE 6-10. Air photo of San Francisco peninsula showing areas of greatest damage during 1906 earthquake (cross-lined areas). Solid line marks landward edge of bay muds (from Nichols, D.R., and Buchanan-Banks, J.M., 1974, Seismic Hazards and Land-Use Planning, U.S. Geol. Survey Circular 690).

(g). Modern high-rise structures in southern California are designed to withstand a horizontal acceleration of about 0.35 g (35% gravity) and single family dwellings about 15% of gravity. The highest horizontal acceleration measured to this date was about 1.05 g in the San Fernando,

California earthquake of 1971. The quake was only a Richter magnitude 6.4 and the duration of strong ground shaking lasted only 12 seconds! However, because of the high accelerations the property damage, mostly to freeway bridges and overpasses, was almost a billion dollars.

Minimizing Hazard. A first step to minimize the hazards of ground shaking might be to pass a geologic hazard zoning act that would limit construction on thick, water-saturated sediments such as the muds marginal to bays like San Francisco Bay. Another action, designed to prevent loss of life, is related to abatement of hazardous structures. This is a difficult problem because of our free enterprise system, but ordinances can be passed to bring older structures up to code and to require removal of dangerous parapets that topple onto sidewalks and streets during earthquakes (Fig. 6-11). Falling debris is a prime killer and is the reason that, if you are indoors

FIGURE 6-11. Unreinforced parapet collapsed on this "convertible" during the Santa Rosa earthquake of 1969 (magnitude 5.7). Loss of life and injury can be severe during moderate quakes from collapse of substandard structures.

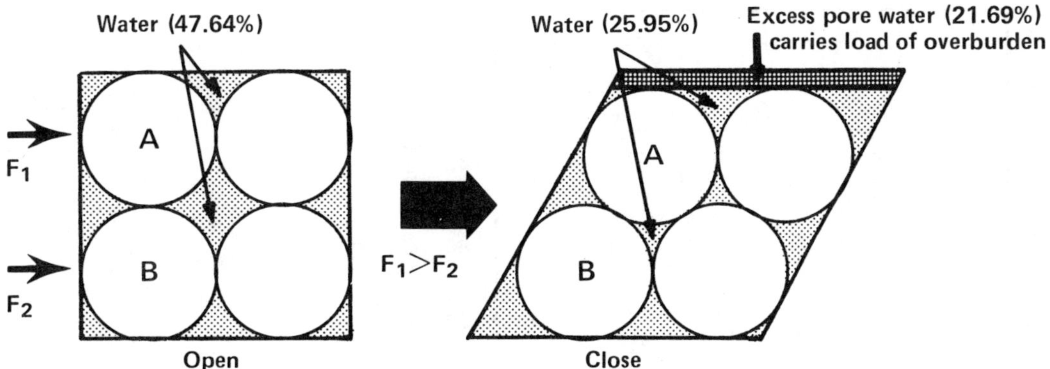

FIGURE 6-12. Liquefaction resulting from unequal acceleration of idealized spheres (representing sand grains) during an earthquake. (After Borchardt, Glenn, and Kennedy, M.P., 1979, Liquefaction Potential in Urban San Diego, Calif. Geology, Calif. Div. Mines and Geology.)

during a quake, you should stay there (see Earthquake Preparedness later in this chapter). Such ordinances have the potential for sharply reducing loss of life during a quake.

LANDSLIDES

Downslope movements of large masses of earth and rock may be triggered by ground vibration during an earthquake. Such failures usually occur on slopes that are weak to begin with and in many cases are predictable by a geologist (see Chapter 8). It is incumbent upon the planner and geologist to identify the potential hazard and to design facilities accordingly.

FOUNDATION FAILURES

Foundation failures involve the loss of strength of incoherent earth materials where flowage is away from structures they support. A common mechanism of failure in water-saturated fine sands is called **liquefaction** and is the result of changes in the packing arrangement of sand grains (Fig. 6-12). Saturated sand in the loose-packing arrangement has pore space volume (porosity) of about 48 percent. During a quake the spheres representing sand grains are not accelerated equally and rearrange into the close-packed configuration with a porosity of about 26 percent. If the excess pore water (about 22 percent) cannot escape it then carries the load of the overburden formerly supported by grain-to-grain contact of the sand. The overburden load on the water then becomes excess pore pressure and pushes the sand grains apart much like quicksand. If this layer is near the surface, buildings and overburden will sink into the liquefied mass, or if there is an open surface on one or more sides, the material will flow laterally. Liquefaction only occurs in clean fine sands, as shown in Figure 6-13. The reason is that a high percentage of silt or clay in the sand will prevent the spheres from rearranging into a closer packing arrangement.

Loss of strength in fine-grained (clay or silt) cohesive materials is another mechanism of foundation failure. Such failures manifest themselves by lateral spreading on flat or nearly flat

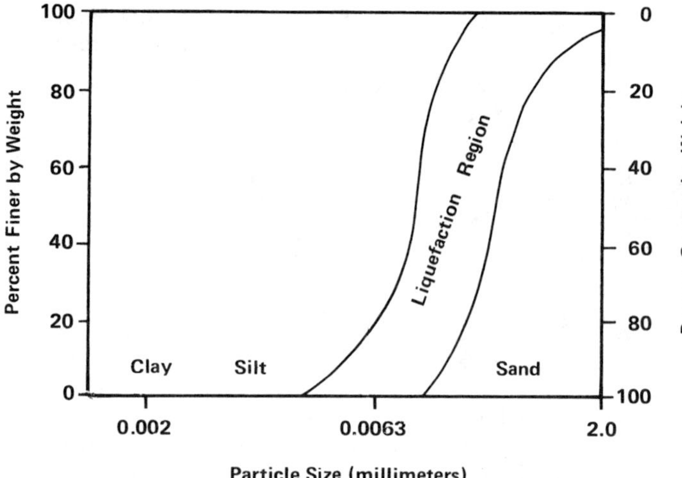

FIGURE 6-13. Graph of particle-size range of sand and silt subject to liquefaction.

surfaces with the development of huge cracks and settlement of massive blocks of earth. A classic example is the lateral spreading brought about by rearrangement of clay particles due to vibrations from heavy artillery explosions at military establishments (Figure 6-14).

Foundation Failure Assessment. Maps showing depth to bedrock and depth to the water table would be useful in assessing liquefaction potential. Preliminary site investigation must include deep bore holes (100 feet ±) where liquefaction potential exists, such as in coastal lagoon and flood-plain areas that have high water tables and clean, sandy sediments. In lieu of detailed studies, maps indicating areas needing further study could be assembled for the planner and public building official.

TSUNAMI AND SEICHE

Tsunami, or **seismic sea waves**, are a potential hazard throughout the Pacific Ocean basin and parts of other oceans. These impulsively generated waves are to be feared more for their uprush onto the land than for the destructive force of the breaking wave itself. Of great use to the planner and engineer are maps showing runup, based upon historic events, indicating the highest levels to which tsunami runup is likely to reach. Hawaii has been hit by scores of tsunami in the last century, many of which have caused catastrophic loss of life. There are sufficient historic data on runup to construct maps such as those shown in Figure 6-15. These maps are for disaster preparedness as well as planning, and they appear in the telephone book so that all can know where runup of seismic sea waves will occur and avoid these areas in event of a tsunami warning.

Seiche is the sloshing of water in closed or partially enclosed basins such as lakes, bays, and harbors. They are forced oscillations and may cause alternating flooding and retreat of waters at certain points, called **antinodes**, in lakes and harbors. Generally, seiches do not pose a catastrophic hazard but they can cause damage in marinas and yacht anchorages because of the rapid rise and fall of the water. Protection against seiche damage is beyond the scope of this manual. Coastal engineers can design harbors to minimize these oscillations.

EARTHQUAKE PREPAREDNESS

It is often said that "the best defense is a good offense." This means that preparation of your physical surroundings before a quake, and proper response during an event, is the best defense against one of nature's most violent hazards. The actual movement of the ground during a quake is seldom the direct cause of death or injury. Most casualties occur because of objects falling from buildings and other structures. Your home is probably the safest place during an earthquake; you should also, however, become familiar with emergency procedures at your place of work or study.

To minimize damage at home or in an apartment or dormitory, anchor bookshelves and other large standing units to the wall. You should know how to turn off utilities, particularly gas, if damaged during a quake. An emergency survival kit should include bottled water, canned

FIGURE 6-14. Quick clay slide of November 12, 1955, at Nicolet, Quebec, Canada. In less than 7 minutes the sliding clay left a hole 600 feet long, 40 feet wide, and 20–30 feet deep.

foods, first-aid supplies, and a battery operated flashlight and radio. The following suggestions are of survival value when an earthquake strikes.

During an Earthquake

1. Don't panic. Stay calm.
2. If indoors, stay there. Get under a heavy desk, or under a doorway, and away from windows or glass.
3. If outdoors, don't enter buildings. Stay in the open, away from buildings, utility wires, trees, and telephone lines.
4. Don't drive. Stop your car and stay inside.
5. Don't use elevators. If inside a high-rise building, stay there and get under a desk. After the shaking has stopped, use the stairs.

After an Earthquake

1. Be aware that there may be aftershocks. If so, continue to take safety precautions.
2. Check for injuries and wear shoes to avoid further injury.
3. Check for unsafe conditions (gas leaks, water leaks, or shorting electrical wires). If any dangers are detected shut off utilities.
4. Turn on battery radio or car radio for emergency bulletins.
5. Don't use candles, matches, or any open flames. Have a flashlight with batteries accessible.
6. Don't use telephones except for emergency.
7. Don't go sightseeing.
8. Don't touch downed power lines or objects touched by downed power lines.

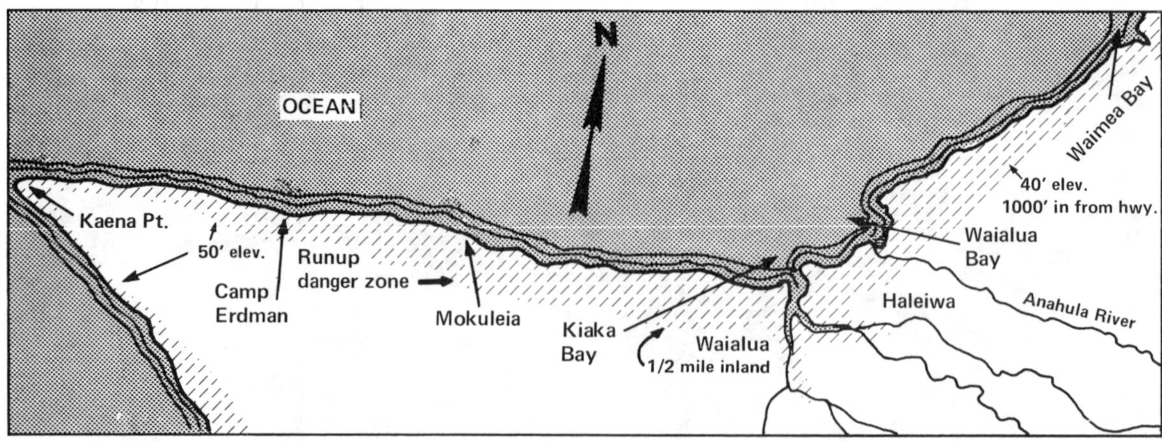

FIGURE 6-15. Tsunami runup on the north shore of Oahu, Hawaii. This map includes the famous board-surfing areas of Waimea and Sunset Beach (from Honolulu Telephone Book).

EXERCISE VI

SEISMIC HAZARDS AND LAND-USE PLANNING

Name ______________________________

1. Sketch a line drawing of a normal, a reverse, and a strike-slip fault.

A B A B

normal reverse strike-slip

2. Name three geomorphic features found along major active fault zones.

   (a) scarp

   (b) bench

   (c) sag pond

3. Study Figure 6-4 and name the compatible cultural elements in the Hayward fault zone.

   underdeveloped area where there are traces of fault.

   What cultural elements pictured should not have been built in this area?

   schools, housing development, + nursery

   Name at least three other uses for the land not pictured in the fault zone.

   ____

4. From the graphs of Figures 6-6 and 6-7, determine the following for a magnitude 8 earthquake (data applicable to western United States only):

   Rupture length of fault 50 miles, ________ km

   Probable displacement or offset 4 feet, 4.3 meters

   Recurrence interval < 1 per 1 years

5. Figure 6-16 is a map showing relative ages of faults in the Sugar House Quadrangle, Utah. Each fault is indicated by a color that identifies the time when the last movement occurred on the fault: red, for which some movement occurred within the last 5,000 years; purple, for which some movement occurred between 5,000 and 3,000,000 years ago; and green, for faults that have been inactive for at least 3 million years. Assume that the recently active fault (red) has the potential for surface breaks within a zone 0.1 mile wide on either side of the fault called a "setback" zone.

   (a) Pencil-in the setback zone on either side of the fault extending it to the area where the fault is dotted and queried (?).

   (b) Is the water tank (WT) in Section 14 within the zone? yes

   (c) Is the Cottonwood School within the zone? no

   (d) If violent ground shaking accompanied an earthquake, where would earthquake-induced rock slides and landslides occur?

   on soft ground

   (e) How would you view a proposal to build a dam at the mouth of Tolcats Canyon that would impound water and create a large reservoir for water storage and recreation? Give geologic reasons for or against the proposed construction.

   should not be done - could cause alternate flooding and retreat of water at antinodes

   (f) Assume that earthquake-induced ground shaking resulted in extensive damage to all buildings within a one-mile radius of the epicenter. The epicenter is where the solid red line and dotted red line join and there is surface fault rupture.

   How many schools would be affected? none

   What utilities would be affected? telephone, sewers, buildings, power lines

   (Note, utilities may not be explicit on the map, but include in your answer such systems as telephone, sewers, and . . . ?)

   What major roads, generally used for emergency vehicles, would be affected?

   Wasatch Blvd., Holladay Blvd.

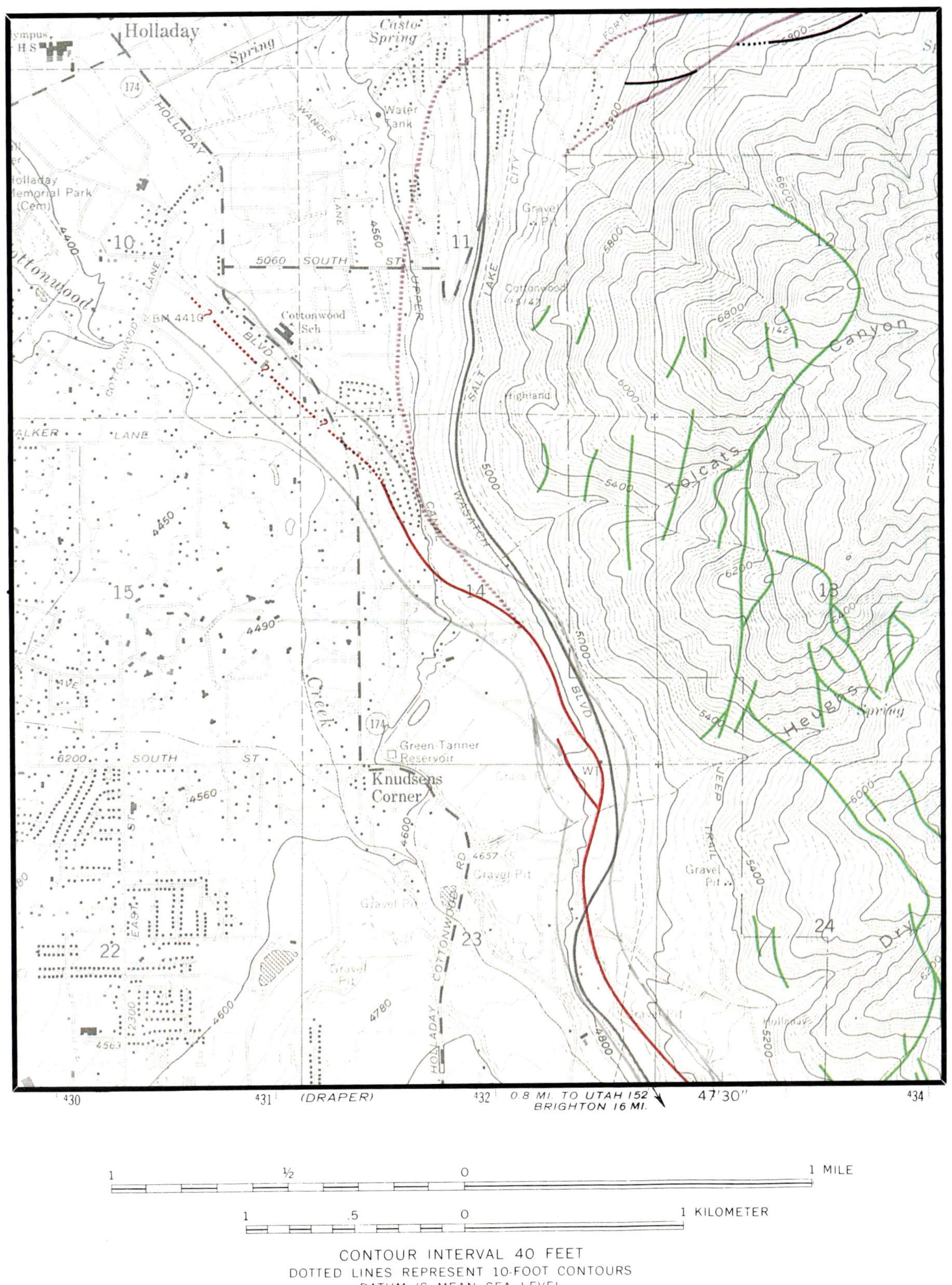

FIGURE 6-16. Ages of faults, Sugar House quadrangle, Salt Lake County, Utah. (Courtesy U.S. Geol. Survey.)

(g) In Section 14, the fault colored red joins another one colored purple. If the active fault produced an earthquake located as in (f) do you think the older fault would be affected and, if so, in what way?

no ______________________________

____

(h) On a sheet of tracing paper prepare a land-use map of the area, basing your plan only upon active faulting and associated ground shaking. For land-use purposes you may classify terrain as (1) unlimited use, (2) limited use without large permanent structures, (3) active recreation and light construction, or (4) passive recreation only, no structures.

6. Figure 6-17 is a simple geological map of a seismically active land area. Which of the earth materials on the map would have the highest seismic response (most shaking)?

aluvium + river deposits ______________________________

Which material would respond least? bedrock ______________________________
Indicate by cross-hatching in ink or pencil the areas of potential liquefaction or lateral spreading.

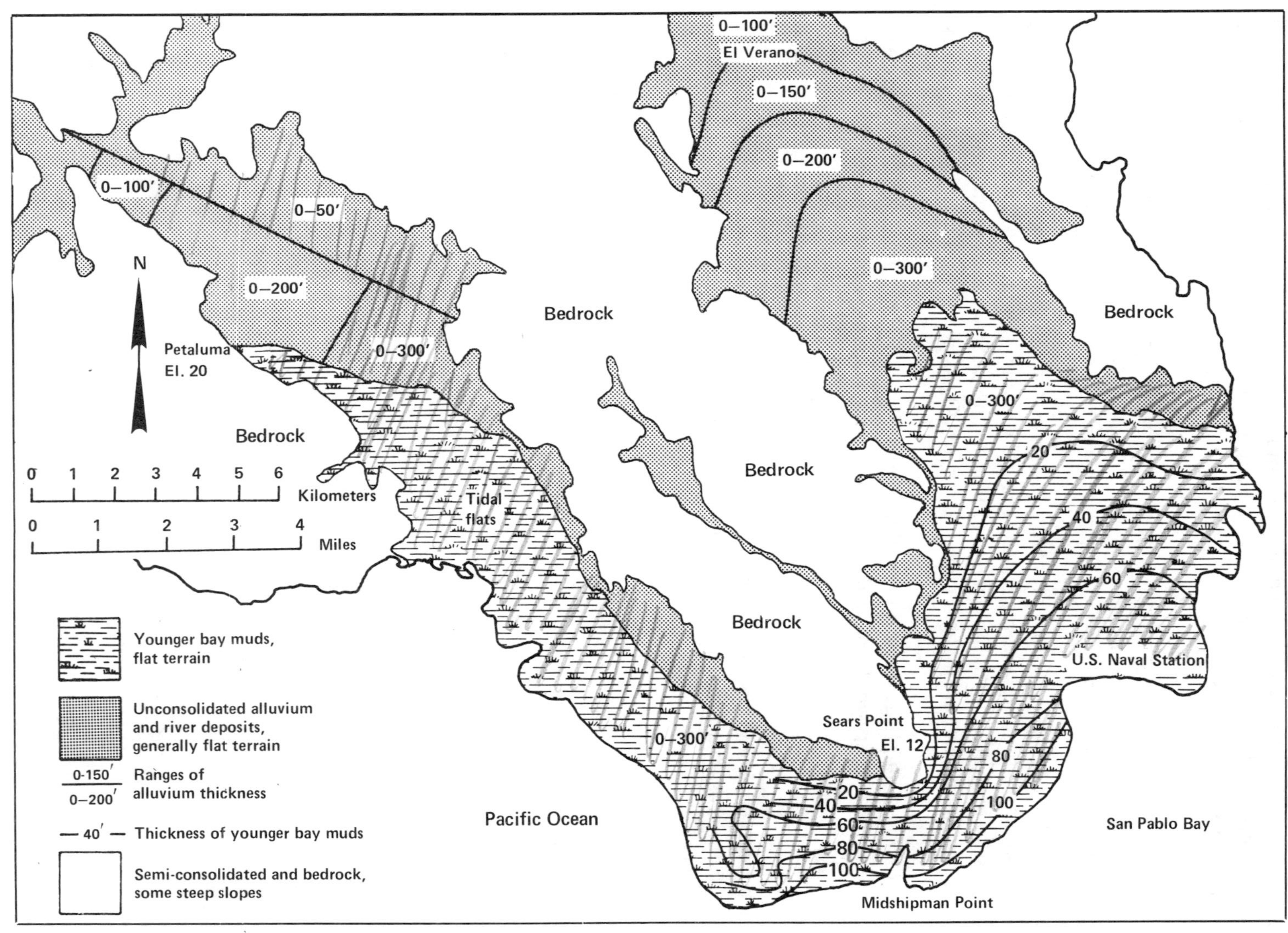

FIGURE 6-17. Relative hazard from seismic shaking in southern Sonoma County, California. (After Greensfelder, Roger, 1974, California Div. Mines and Geology Special Report 120.)

Chapter 7

Volcanoes — An Urban Hazard

On May 18, 1980, Mount St. Helens in Washington exploded with atomic-bomb violence ending a quiet period of 123 years. The laterial blast devasted 300 square miles of timber and recreational area. It caused flooding and debris flows in the Toutle and Cowiltz Rivers watersheds; debris, logs, and sediments in these rivers were carried to the Columbia River estuary at Portland to disrupt commerce. In the wake of the lateral blast and subsequent pyroclastic flows and ash falls, 60 people lay dead or missing and property damage was estimated at 2 to 3 billion dollars. Had the catastrophy occurred at a time other than Sunday morning, no less than 700 lumberjacks and loggers would have been in the area subject to the fatal wilting blast of hot air, steam and ash!

Following the lateral blast that produced the fatalities and property damage, a vertical plume of ash rose 20 km into the stratosphere and was carried by the prevailing winds into western Washington, Montana, and Oregon. By late afternoon, one-half inch of ash had fallen at Spokane creating a colossal cleanup problem. In a few days, ash was noted in New York State and within a matter of weeks ash and dust were carried around the world. The most interesting facet of the Mount St. Helens' scenario is that it was predicted by geologists in great detail, only the violence of the explosion was underestimated!*

Mount St. Helens is one of a number of aligned volcanic cones that compose the Cascade Range (Fig. 7-1). Many of the volcanoes, such as Baker, Hood, Ranier, and St. Helens, record a violent eruptive history that goes back about one million years. The alignment and youthfulness of the cones has been the subject of much scientific speculation; however, plate tectonics offers a scientifically supportable theory for their origin. Mount St. Helens is the youngest (37,000 years ±) and most westerly of the chain; thus it is no accident that it is most recently active. Offshore, to the west, lies the Juan de Fuca ridge (spreading center) and subduction zone. The new sea floor created at the ridge spreads easterly to be subducted beneath the western margin of the continent. The subducted slab remelts and rises as a plume beneath Mount St. Helens (Fig. 7-2).

---

*Crandell, D.R., and Mullineaux, D.R., 1978, Potential hazards from future eruptions of Mount St. Helens volcano, Washington: U.S. Geological Survey Bulletin 1383-C, 26 p. This publication preceded the eruption by 2 years.

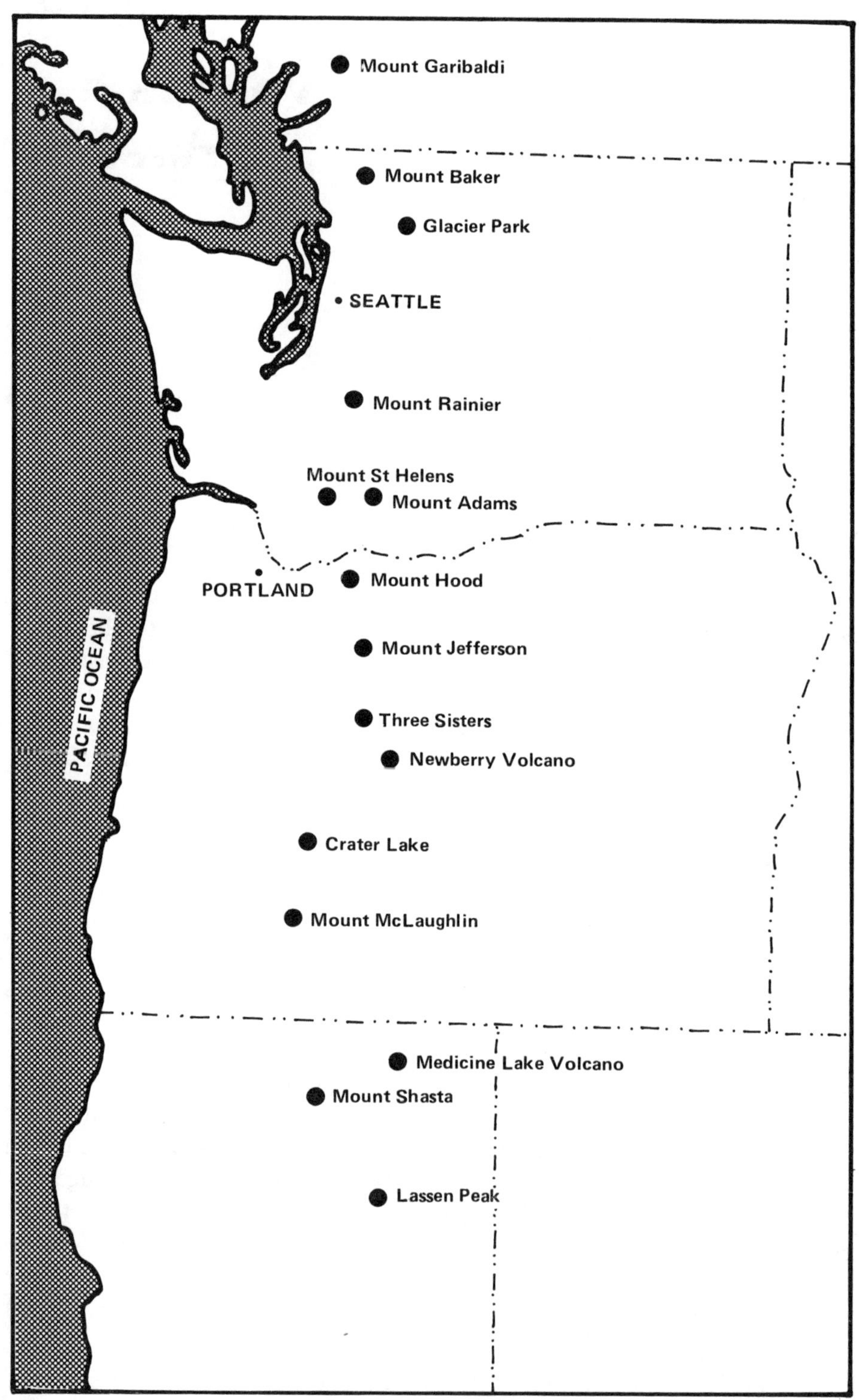

FIGURE 7-1. Location of Cascade Range volcanoes.

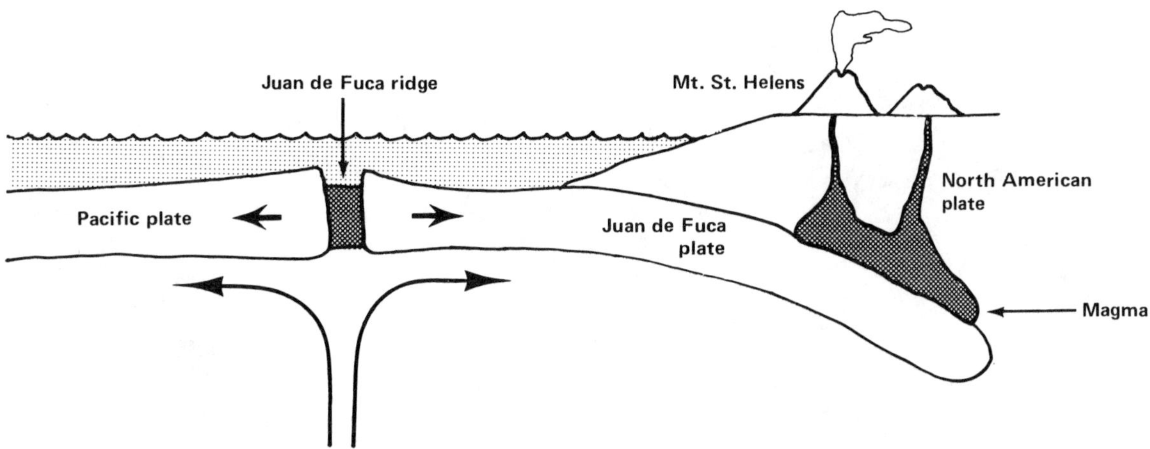

FIGURE 7-2. Plate tectonic framework for the eruption of Mount St. Helens. Spreading from the Juan de Fuca ridge and subduction at the continental margin explain many of the active volcanoes of the Cascade Range.

CLASSIFICATION OF VOLCANIC ACTIVITY

Active volcanoes are distributed around the Pacific Ocean basin, the so-called "Ring of Fire"; along the mid-ocean ridge systems; and through a Mediterranean-Trans-Asian Zone (Fig. 7-3). It is no accident that the distribution of active volcanoes follows closely plate boundaries and major earthquake belts.

Volcanic activity may be classified as either explosive or quiet. **Explosive** activity is generally confined to continental margins or island arcs bordering the continents. Volcanoes of the Cascade Range in the United States, in Central America, in Japan, and in the Philippines are examples. Figure 7-4 is a photograph of Cerro Negro in Nicaragua. **Quiet** activity, on the other hand, is generally found in oceanic volcanoes and along the mid-ocean ridges and rises such as those of Iceland and Hawaii.

Explosive volcanoes build a cone that has the classic concave upward profile so characteristic of famous ones like Fujiyama in Japan and Mayon in the Philippines. The lower slopes are composed of lava flows that assume low angles, whereas the upper slopes are composed of cinders and ash that have a higher natural angle of repose. As a result these are known as **composite cones**. Destructive effects of eruptions, other than rare explosive events of catastrophic scale, would probably be limited to areas within a few tens of kilometers down valleys of downwind (Mullineaux, D.R., 1976, Volcanic Hazards, U.S. Geological Survey Misc. Field Studies Map MF786).

Remnants of once-impressive composite cones that exploded or collapsed are to be found at Crater lake in Oregon and Mt. Katmai in Alaska. It is estimated that 14 cubic miles of the former cone, which once occupied the area around Crater Lake, was either blown away or collapsed into the underlying magma chamber. This undoubtedly destructive event has been dated by carbon-14 at about 6,600 years ago. Similarly, Mt. Katmai on the Cook Inlet in Alaska exploded in 1912 leaving only a huge depression with thousands of steam vents known as **fumeroles**. The steam

FIGURE 7-3. Worldwide distribution of (a) volcanic eruptions and (b) major earthquakes (Richter magnitude 7.0 and above), 1968 to 1974. (Center for Short-Lived Phenomena.)

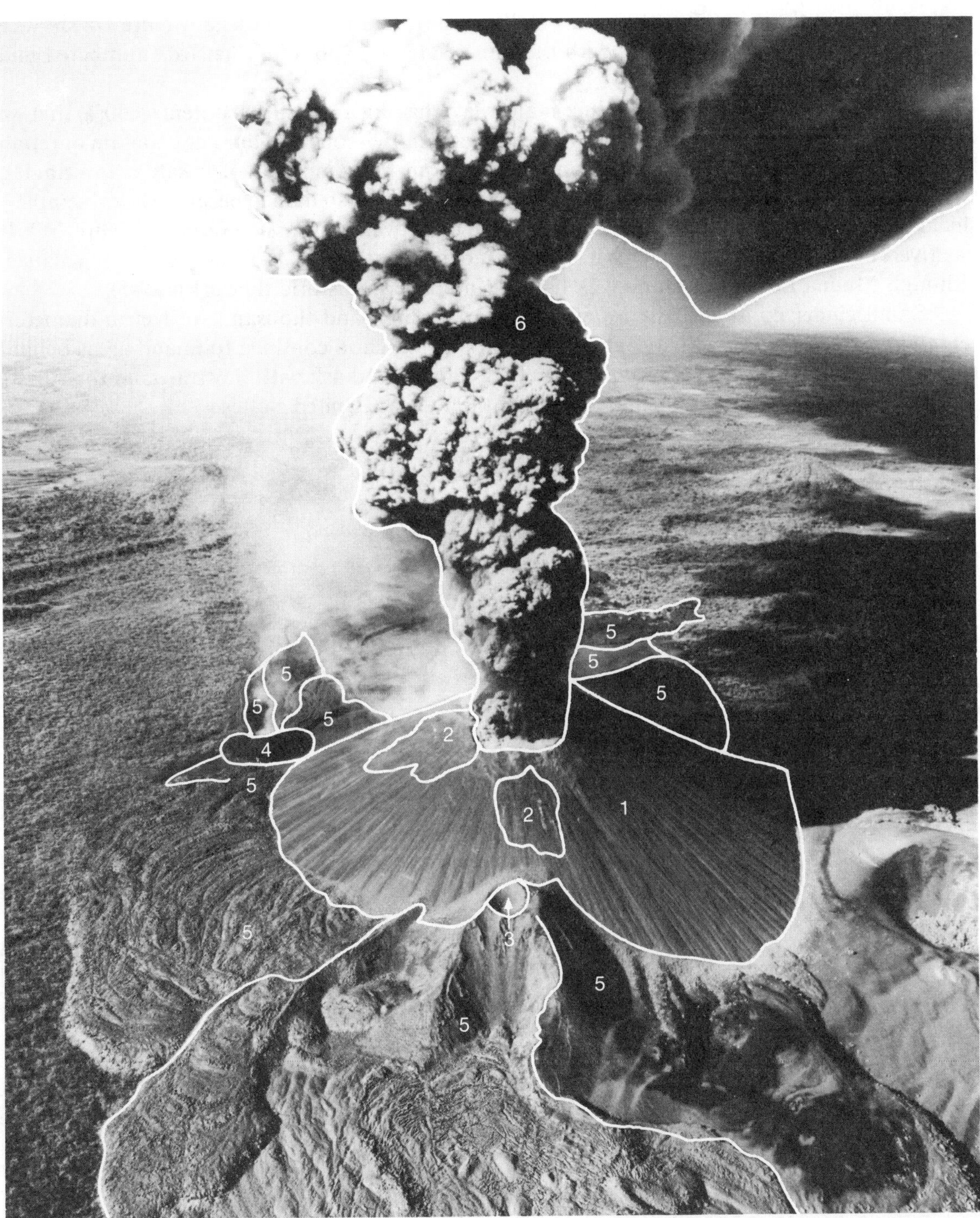

FIGURE 7-4. Eruption of Cerro Negro, an active composite cone volcano in Nicaragua. (Courtesy California Division of Mines and Geology.) Legend: 1. Cerro Negro composite cone; 2. smoke from recently ejected materials; 3. crater of an active, small parasitic volcano on the flanks of Cerro Negro; 4. active parasitic vent; 5. separate lava flows that have emanated from parasitic vent at (4); 6. cloud of pyroclastic debris that has been ejected from the volcano and will be deposited downwind.

that comes from the vents gives the name "Valley of Ten Thousand Smokes." Figure 7-4 shows a composite volcanic cone in action, with the immediate and subsidiary hazards numbered and described.

Oceanic or quiet volcanoes produce mainly lava having a low silica content (<50%) that we call basalt. Because of its low silica content, the lava is not viscous and does not contain or retain large volumes of gases. Thus these eruptions build up huge mounds of basalt at low angles, usually less than 12 degrees, known as **shield cones**. They are so named because they resemble a shield with the convex side upward. The big island of Hawaii is a classic example (Figure 7-5). It has five known vents, all named, that operate independently. The most active vent is Kilauea, although Mauna Loa and Mauna Kea have been active in historic times (Fig. 7-6).

Small **cinder cones**, usually a few hundred feet high and thousands of feet in diameter, represent a third type of activity. They are local phenomenon confined to inland areas behind leading edges of plates. They are built entirely of cinders and ash with flows around their base. There are thousands of young cinder cones in the western United States.

FIGURE 7-5. Kilauea volcano in foreground with shield-shaped cone of Mauna Loa in background. (U.S. Geological Survey photo.)

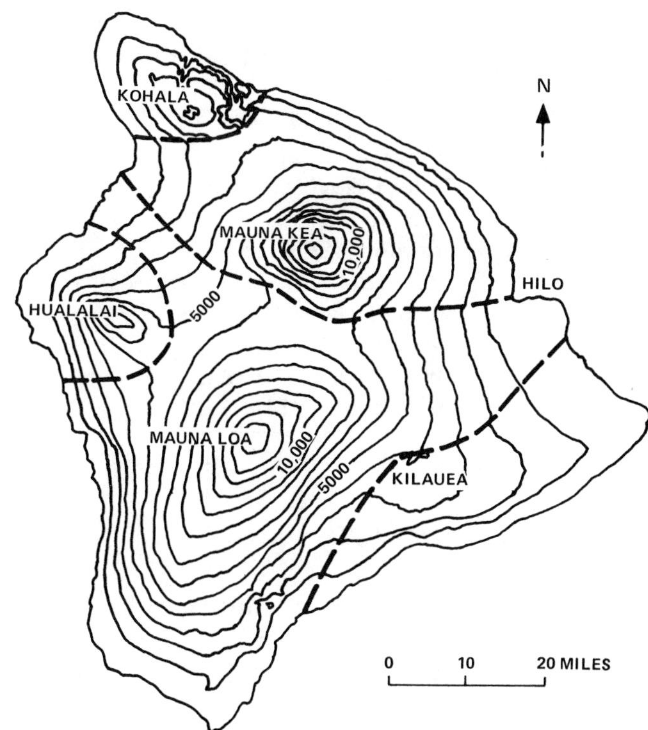

FIGURE 7-6. Separate volcanoes on the island of Hawaii. Note location of city of Hilo in relationship to the vents of Mauna Loa and Mauna Kea.

MAGNITUDE OF VOLCANIC ERUPTIONS

The measure of the magnitude of a volcanic eruption is based upon the amount of material ejected during the eruption (Table 7-1). The highest rank is a magnitude 1, which is an eruption greater than 1 cubic kilometer (1 billion cubic meters) of new material. A few famous eruptions are listed in Table 7-2, with the estimated or measured volumes or eruptive material in decreasing order. It is immediately obvious that the great catastrophy of 1980 in the Pacific northwest was really an average large eruption in the context of the recent volcanic history of the Earth. However, the impact of this type of volcanic eruption on the local culture, topography, vegetation, and wildlife can be significant and will be explored in this chapter.

TABLE 7-1
MAGNITUDE OF VOLCANIC ERUPTIONS

Magnitude	Volume of Ejected Material (cubic meters)	Magnitude	Volume of Ejected Material (cubic meters)
1	10^9 (1 billion)	5	10^6–10^5
2	10^9–10^8	6	10^5–10^4
3	10^8–10^7	7	10^4–10^3
4	10^7–10^6	8	Less than 1,000

TABLE 7-2
SOME FAMOUS ERUPTIONS

Volcano	Volume New Material (billion cubic meters)
Crater Lake (Mt. Mazama)	42
Tambora (1915)	25
Krakatoa (1883)	18
Mt. St. Helens (3,000 years ago)	8.0
Mt. St. Helens (450 years ago)	2.6
Vesuvius (79 A.D.)	2.6
Mt. St. Helens (1980)	1.6–2.0
Mt. Lassen (1914)	1.0

VOLCANIC PRODUCTS

Volcanic ejecta may be classified as either **lava** or **pyroclastics.** Lava is molten material at the surface of the Earth, whereas pyroclastics are broken fragments blown from the vent of the volcano. Pyroclastics are classified according to size as follows:

block .... old cold material greater than 32 mm
bomb .... new hot material greater than 32 mm
lapilli (cinders) .... 32 mm to 4 mm
ash .... 4 mm to 1/4 mm
dust .... less than 1/4 mm

Pyroclastic material usually falls from the atmosphere (ashfalls) and is crudely stratified and subject to the prevailing winds. Lavas, on the other hand, flow downslope and follow natural watercourses and valleys. Flows move slowly and may have a blocky surface. A rough blocky "clinker" is developed on both the top and bottom of the flow. This type of lava is called **aa** (ah-ah) in Hawaii. Lava that flows rapidly in thin sheets has a smooth ropy surface and is called **pahoehoe** (pa-hoy-hoy). Both forms of lava may occur in composite cones and cinder cones (Fig. 7-7).

Corrosive gases that usually form hydrochloric acid or sulfuric acid with moisture in the atmosphere are common products of volcanoes. Huge volumes of water vapor also are produced and may result in intense rainfall during or following an eruption.

GEOLOGIC HAZARDS

Risk from volcanic hazards decreases with distance from the source of the eruption. Lava flows are nearly uniformly destructive to their terminus; however, the probability of damage from them decreases with distance from the vent or fissure. Ashfalls also become less destructive and less hazardous with distance from the source. As one goes farther from the source, the thickness of ash decreases, as does the probability of being inundated.

FIGURE 7-7a. Lava from Ulu crater on the southeast rift zone of Kilauea volcano.

FIGURE 7-7b. Pahoehoe lava from same area.

Lava Flows. Damage from lava flows is usually confined to those areas downslope from the erupting vent. Whole communities have been buried by slowly moving lava flows, such as Paricutin in Mexico and parts of Vestmannaeyjar on Heimaey Island, Iceland (Fig. 7-8). The course of the lava flow will follow existing drainage and can be predicted once the flow is established.

Ashfalls. Ashfalls have been known to bury agricultural lands and urban areas in layered cinders, ash, and dust. Although quickly broken down into rich soil, such ash layers have a devastating short-term effect. In some cases, where accompanied by heavy rains, the ash is so heavy that it causes buildings to collapse. Such was the case at the Roman resort city of Pompeii on the Bay of Naples. The pyroclastic eruptions of 79 A.D. were so voluminous that the city was totally covered as if by a snow blizzard. Many victims were buried in their homes because of difficulty in traversing the sodden ash on highways leaving the city.

Ashfalls move downwind rather than downslope from an eruption center. Prevailing wind direction determines the probability of a site being affected by an ashfall, and the velocity of the wind determines the distance from the site that the airborne material will travel.

Ash flows. Ash flows are hot masses of gas, pumice, and pyroclastics that, because of their density, move as a dust storm along the ground with tremendous velocities. Thus everything in their path is burned, boiled, or suffocated depending upon the nature of the flow. In 1902, an ash flow from Mt. Pele, in Martinique in French West Indies, roared onto the city of St. Pierre and killed all of its 40,000 inhabitants save one. Even seamen aboard ships in the harbor were suffocated or burned to death. These blasts are unique and require a special set of circumstances for them to occur. The deposits that result are very distinctive and we can recognize them in ancient rocks. They are called "welded tuffs" because the mineral grains are flattened and welded together from the intense heat. In addition, we can recognize base surge deposits characteristic of high-velocity particle-laden flows. Such deposits have been identified at Yellowstone Park, at Long Valley, California, and in the Jemez Mountains of New Mexico. At Long Valley, the Bishop Tuff has a volume of 38 cubic miles and ash layers have been recognized 400 miles away.

Not all ash flows are hot enough to cause burning of objects in their path. At Mount St. Helens, the pyroclastic eruption plume, consisting not only of ash, but also of lapilli, dust, and large blocks, was not hot enough to burn logs or timber in its path.

Tsunami. Tsunami, or seismic sea waves, may result from large submarine volcanic eruptions and also from fault movement on the ocean floor. A 100-foot-high wave was generated by the explosion of Krakatoa in 1883. This wave struck the southwest coasts of Java and Sumatra and reportedly killed 30,000 people. Submarine eruptions of this magnitude are rare and present a statistically low hazard.

Floods and Mudflows. Melting of glacier ice on large composite cones in northern latitudes presents a real hazard, particularly in heavily populated areas such as Oregon and Washington and parts of Alaska. As activity and heat flow increase, ice barriers may fail or natural dams may be breached and catastrophic floods may occur downstream. At present, there is a real concern in the Cascade Range of Oregon and Washington that renewed melting activity may create such flood conditions. In addition to any floods caused by melting ice, the large volumes of water expelled by volcanoes may condense and produce heavy rains. Dense, viscous mixtures of fine-grained volcanic materials may also develop and cause extensive mudflow damage during eruptions.

FIGURE 7-8. Volcanoes are truly an urban hazard on the island of Heimaey in Iceland. Eldfell volcano erupted in 1973 and lava and pyroclastic material inundated the fishing village Vestmannaeyjar on the island. It also marked man's most successful attempt to control laval flows by cooling them with seawater (U.S. Department of the Interior, Geology Survey).

MOUNT ST. HELENS — A CASE HISTORY

All of the hazards discussed in the previous section, except lava flows, occurred during or after the Mount St. Helens sequence. On Sunday, May 18, 1980, at 8:30 a.m., a magnitude 5.0 earthquake triggered a massive landslide on the north side of the volcano. This landslide, probably one of the largest ever witnessed by man, moved down the gentle slope of the mountain for a distance of 8 km. About half the mass slid into Spirit Lake north of the volcano and the remainder was diverted by a ridge into the North fork of the Toutle River. Almost simultaneously a lateral blast of incandescent gas and ash of hurricane proportions blew down trees, stripped bark, and caused most of the fatalities. This blast had an estimated velocity of 180 km per hour and temperatures on the order of 200° C, hot enough to scorch but not char trees 30 km from the vent (Fig. 7-9). Within minutes of the lateral blast, a vertical cloud of ash and dust developed that reached a height of 20 km (70,000 ft) (Fig. 7-10).This cloud traveled northeast from St. Helens and was deposited in the pattern shown on Figure 7-11. During the continued vertical venting of the volcano, pyroclastic flows moved down the Toutle River valley and raised its level by more than 90 m (Fig. 7-12). Finally, within a few hours of the initial eruptions, debris flows and floods from the displaced lake water and melting ice flowed down the North and South forks of the Toutle River destroying logging camps and private homes. The total area affected was about 500 km^2 (300 mi^2) and total property loss about $2.7 million. Figure 7-13 shows the areas impacted by the blast and subsequent floods of the May 18th eruption.

PROBABILITY AND PREDICTION

Just where future volcanic eruptions will occur can be predicted with fair accuracy because most eruptions are confined to large central vent volcanoes (Mullineaux, D.R., 1976, Volcanic Hazards, U.S. Geological Survey Misc. Field Studies Map MF786). Some eruptions occur at widely scattered vents or at new ones; these cannot be predicted far in advance with the same accuracy. However, they are usually isolated eruptions and are more likely to occur in some areas than in others. The eruption of Paricutin in Mexico is a good example. It is the only new volcano in North America in historic times and erupted through a farmer's field, grew to several hundred feet in size, and then stopped its activity in a matter of a few months.

Just when eruptions will occur is more difficult to determine. Except in areas of almost continuous activity, such as Kilauea in Hawaii, we can only estimate statistical recurrence based on historical records and geological studies. The bigger the event, the less likely it is to occur, just as in earthquakes. For instance, in the Cascade Range of California, Oregon, and Washington, moderate ashfall eruptions may occur on the order of once every 1,000 to 2,000 years, and very large eruptions are estimated to occur no more than once every 10,000 years. Certainly not pinpoint accuracy. Cataclysmic eruptions (explosive ashflows) have occurred at least three times in the western United States in the last 2 million years. These are recorded in the thick welded tuffs in Yellowstone Park, Long Valley, California, and the Jemez Mountains, New Mexico.

MITIGATING THE EFFECTS OF VOLCANOES

Lava flows have been dammed, diverted, and bombed, with some success. Eruptions from the east rift zone of Kilauea have been diverted around arable lands and it is anticipated that this

FIGURE 7-9. Mount St. Helens blow-down area. Many trees are in excess of 100 feet long (U.S. Geological Survey).

FIGURE 7-10. Eruption of ash and dust into the stratosphere from Mount St. Helens, May 18, 1980 (U.S. Geological Survey).

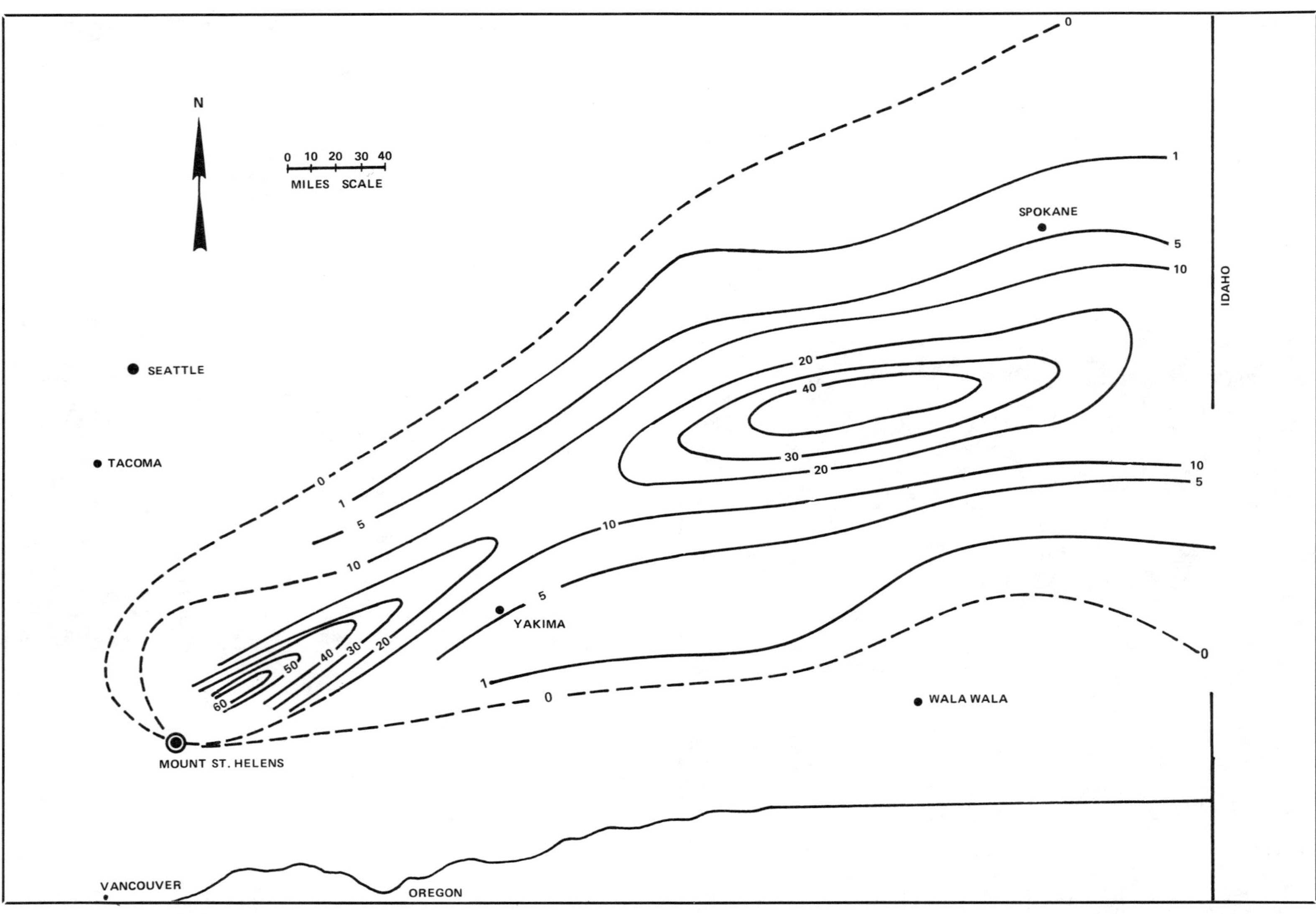

FIGURE 7-11. Isopach map showing distribution of ash from Mount St. Helens major eruption, May 18, 1980. Thicknesses in millimeters not corrected for compaction. (Compiled by A.M. Sarna-Wojcicki and G. McCoy.)

FIGURE 7-12. Debris flows fill the valley of the North Fork of the Toutle River to a depth of 200 feet. Photo taken June 4, 1980 (U.S. Geological Survey).

kind of action will be taken with future flows (Fig. 7-14). Topography is the determining factor because with diversion the lava must be moved in a safe alternate direction. Spraying flow fronts with high-volume jets of water has proved effective in slowing and even checking the flow of lava in both Hawaii and Iceland and will undoubtedly be used more in the future (Fig. 7-15).

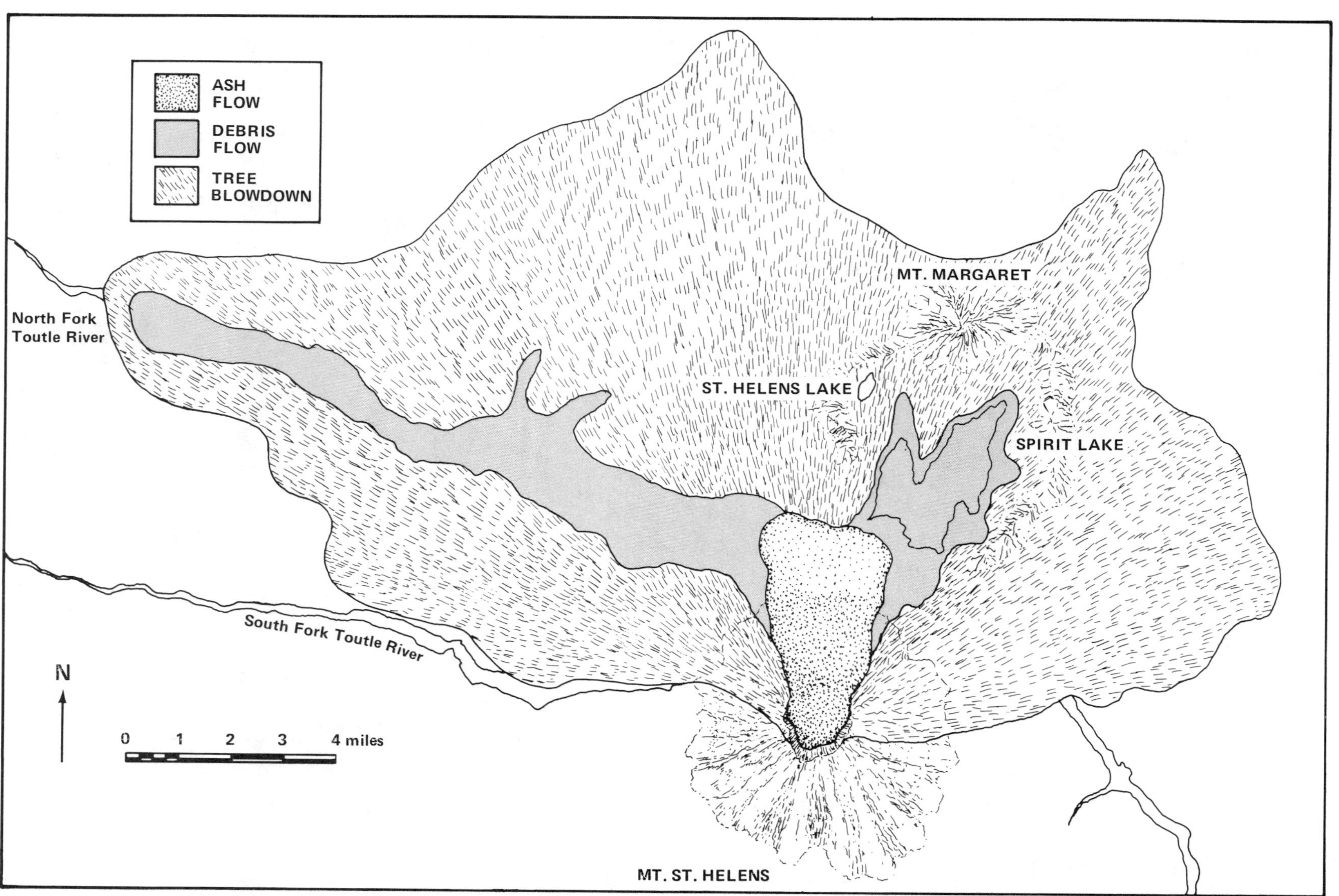

FIGURE 7-13. Approximate limits of damage and distribution of ejected materials from major eruption of Mount St. Helens.

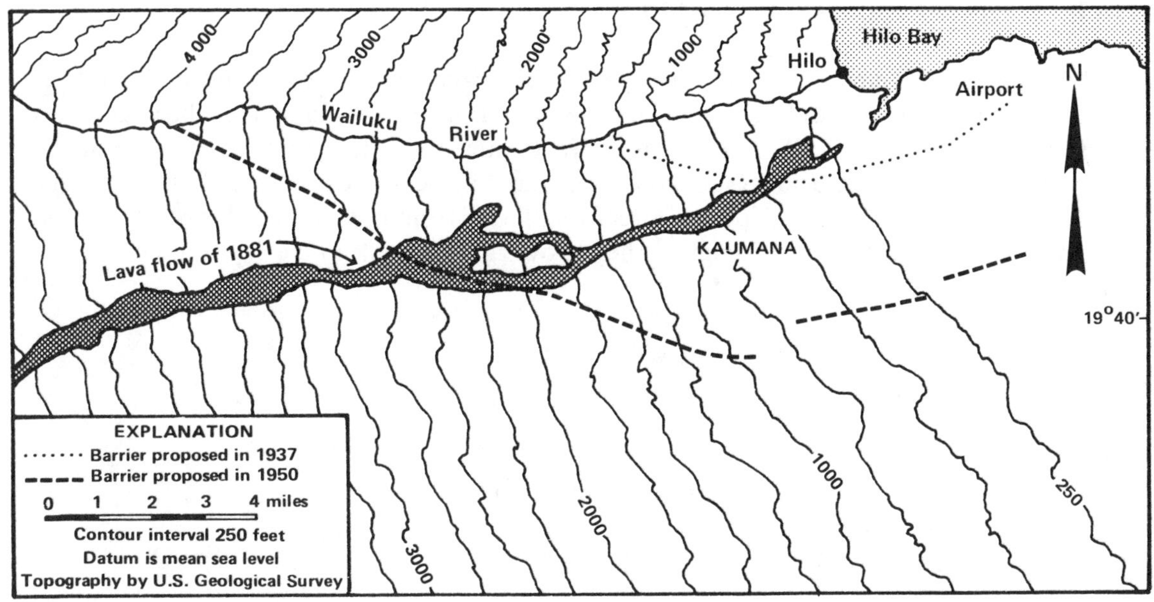

FIGURE 7-14. Map showing Hilo, Hawaii, and the lava flow of 1881. Barriers are proposed to divert future flows. (After MacDonald, G.A., and Abbot, A.T., 1970, Volcanoes in the Sea, Honolulu: Univ. of Hawaii Press, © 1970.)

FIGURE 7-15. Cooling and slowing lava flows by spraying with seawater, island of Heimaey, Iceland. This 1973 eruption was one of the most successfully controlled by cooling and diverting lava flows (Department of Interior, U.S. Geological Survey).

WHAT TO DO WHEN THE ASH FALLS

Volcanic "ash" is not ash at all; rather, it is pulverized rock, cinders, and pumice. Fresh volcanic ash may be harsh and gritty and have an acrid or unpleasant smell. The shape of individual ash particles varies. Although gases are usually too dilute to endanger the normal person, the combination of acidic (sulfurous) gases and ash could cause lung damage to small infants, the very old and infirm, or people already suffering from respiratory illness. The Federal Emergency Management Agency suggests the following when volcanic ash is falling:

IN GENERAL

Don't panic, stay calm.
Stay indoors.
If outside, seek shelter.
Use a mask or moistened handkerchief to aid breathing.
If at work, go home before ash falls.
If ash is falling, stay at work until heavy ash has settled.

A more detailed course of action is suggested when at home or in your auto.

AT HOME

Close doors and windows.
Place damp towels around openings to the outside.
Do not run exhaust fans or clothes dryers.
Remove ash from low-pitched roofs.
Keep refrigerator closed.
Dust often, use vacuum cleaner rather than dust cloths (ash will scratch).

IN AUTO

Drive slowly if you must drive.
Use windshield washers and wipers.
Change air filter every 50 to 100 miles in heavy dust. Filter should not be changed until you notice a loss of power.
Visibility may be sharply decreased, so drive accordingly.

Another federal agency, which shall remain unnamed, advises, "When there is an eruption do not run toward the volcano." Good advice I would say!

EXERCISE VII

VOLCANOES AS AN URBAN HAZARD

Name ________________________________

Study the map and table that describe volcanic activity and hazards in the United States (Fig. 7-16, Table 7-3). This is the type of geological input important for land-use planning.

1. What cities would be subjected to ash inundation from large eruptions? Don't limit yourself to just the cities shown on the map.

   ____

   ____

   What are the specific hazards to life and property within this inner hazard zone?

   ____

2. What major cities or large towns are to be found in the outer hazard zone subjected to ash inundation from very large eruptions? Remember, a "very large" eruption would heavily impact those cities in the inner hazard zone as well.

   ____

   ____

   What are the hazards to be expected in the inner zone from a very large eruption?

   ____

   ____

   What would you expect the impact to be in the outer zone from the same eruption?

   ____

   ____

3. Now study the distribution of ash from the May, 1980, eruption of Mt. St. Helens. How did the ashfall from this eruption compare to the predicted pattern?

   ____

   ____

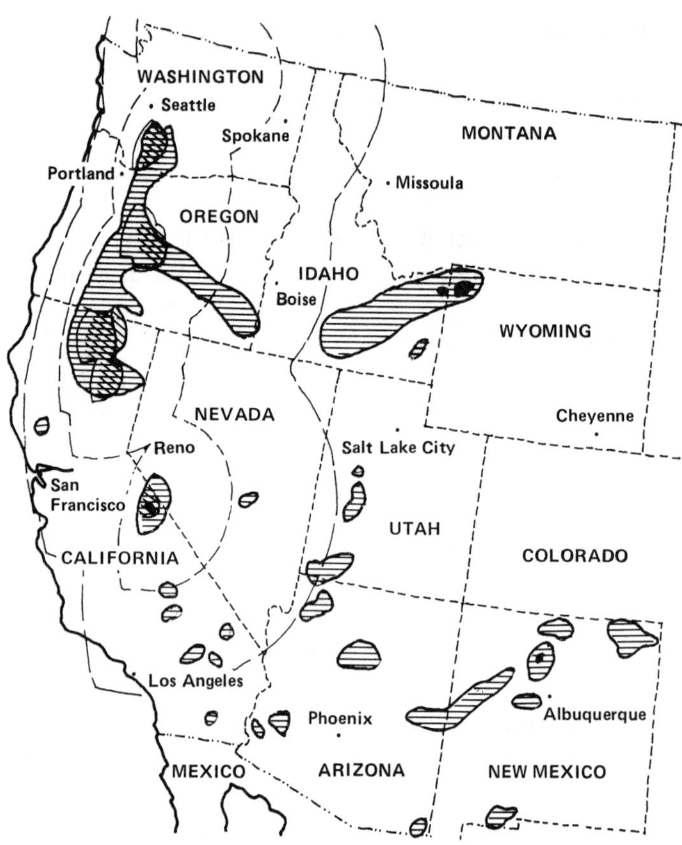

FIGURE 7-16. Map showing volcanic hazard zones in western United States. Solid black zones are volcanic vent areas that had one or more violent eruptions within the last 2 million years. Horizontally lined areas are subject to lava flows and small volumes of ash from groups of vents. Diagonally lined areas are subject to ashfall from explosive volcanoes. The inner dashed line encloses area subject to 5 cm or more of ash from a large eruption. Outer dashed line encloses area subject to 5 cm or more of ash from a very large eruption, like Crater Lake about 6000 years ago. (After D. R. Mullineaux, 1976, Volcanic Hazards, U.S. Geological Survey Misc. Field Studies Map MF786.)

4. Name as many active volcanoes in California, Oregon, and Washington as you can within the inner diagonally lined areas of Figure 7-16 (indicating ashfall from nearby relatively active and explosive volcanoes).

5. What volcanic hazards are to be found in Arizona, New Mexico, Utah, and Idaho?

TABLE 7-3

DESCRIPTION OF VOLCANIC HAZARDS IN THE UNITED STATES

	Lava Flows	Hot Avalanches, Mudflows, Floods	Volcanic Ash and Gases
Origin	Results from quiet eruptions of lava.	Hot avalanches can be caused directly by eruption of fragments of molten or hot, solid rock; mudflows and floods commonly result from eruption of hot material onto snow and ice.	Produced by explosion of fragments and gas into air; materials can then be carried great distances by wind. Gases alone may issue from vents without explosions.
Effects	Land and objects near vent subject to burial and destruction.	Land and objects subject to burning, burial, impact damage, and inundation by water.	Land and objects near vent are subject to blast effects, burial, and infiltration by abrasive rock particles and gases.
Frequency (48 states)	Several small flows per century. Large flows greater than 100 km^2 probably occur about once every 1000 years.	Probably one to several events per century caused by eruptions of relatively "active" volcanoes. Probably one event in 1000 years caused by eruption of relatively "inactive" volcanoes.	Significant eruptions expected once every 100 years. Large eruptions expected once every 1000–5000 years. Very large eruptions probably occur only once every 10,000 years.
Severity	To people, low; to property, high.	Risk can be high to people and property because avalanches and mudflows can originate suddenly and travel at high speeds. Risk decreases away from valley.	Moderate risk to both people and property near erupting volcano; decreases gradually downwind to very low risk.
Predictability	Relatively reliable at large central-vent volcanoes because flows are erupted repeatedly at the same site. Unreliable elsewhere.	Relatively reliable because most originate at central-vent volcanoes and are restricted to valleys leading from them.	Moderately reliable. Ash originates mostly at central-vent volcanoes; its distribution depends mainly on winds.
Size of Area	Relatively small to moderate; large flows cover a few hundred kilometers square.	Confined mostly to floors of valleys and basins that lead to volcanoes. Large, snow-covered volcanoes and those that erupt explosively are principal source of hazard. May extend down-valley many tens of kilometers.	All areas toward which wind blows from potentially active volcanoes are susceptible. Even moderate eruptions could significantly affect thousands of square kilometers.

6. Why is there essentially no risk from volcanic hazards in the central or eastern United States? ______________________________

____

7. List the statistical frequency of the following volcanic events in the contiguous United States (48 states):

Lava flows ______________; hot avalanches, mudflows, and floods __________;

volcanic ash and noxious gases ______________________________.

8. What should you do during an ashfall?

At home ______________________________

____

____

In your car ______________________________

____

____

9. Compare the short- and long-term environmental impact of a volcanic eruption like Mount St. Helens to an earthquake of magnitude 6.5 in an urban area (the impact is on culture, topography, wildlife, vegetation, economy of the area, and the like).

Impact	**Earthquake**	**Volcanic Eruption**
Long-term	______________	______________
	______________	______________
	______________	______________
Short-term	______________	______________
	______________	______________
	______________	______________

Chapter 8
Mass Wasting and Landslides

Slope movements, including landslides, are one of the most costly geologic hazards in the United States. A recent estimate of total property loss due to slope-movements is in excess of $1 billion. The damage can be, and has been reduced through the cooperative efforts of earth scientists, engineers, and public officials. Figure 8-1 is a map of landslide potential in the western United States, which uses rock type and topography to establish high, medium, and low potential for landslide.

Mass Wasting is a general term for a variety of processes by which large volumes of earth materials move downslope by gravity, from the almost imperceptible movements of rock glaciers and soil creep, to awesome, high-velocity rock and snow avalanches. The Earth's surface is irregular and takes the form of topographic highs and lows. This uneven surface topography is not in equilibrium; it is always moving toward becoming completely level. Mass wasting, then, occurs when earth materials move from high points to lower ones.

CLASSIFICATION OF LANDSLIDES

The classification used here was developed by the Highway Research Board of the National Research Council in 1967. It was revised in 1978 (Varnes, David J., 1978, Slope Movement Types and Processes, Natl. Acad. of Sciences, Natl. Research Council Highway Research Board Spec. Report 76, pp. 12–33) and is widely used by engineers and geologists. The basis of the classification is: (1) the type of earth material involved, such as rock, debris (coarse soil), or earth (fine-grained soil) and (2) the type of movement, such as slides, falls, or flows (Table 8-1). We have rock falls and slides; debris falls and slides; and a variety of flows. Flows are divided on the basis of moisture content and rapidity of flow. This division results in a continuum of flows from slow-moving dry rock flows to high-velocity water-saturated debris flows. However, general terms like "landslide" and "mud flow" will continue to be used because they are easily understood.

Landslides are an important form of mass movement and will be the major consideration in this chapter. Note that slides are classified on the basis of the geometry of the slide plane and materials involved. Failure along a planar surface, such as a bedding plane or joint, is

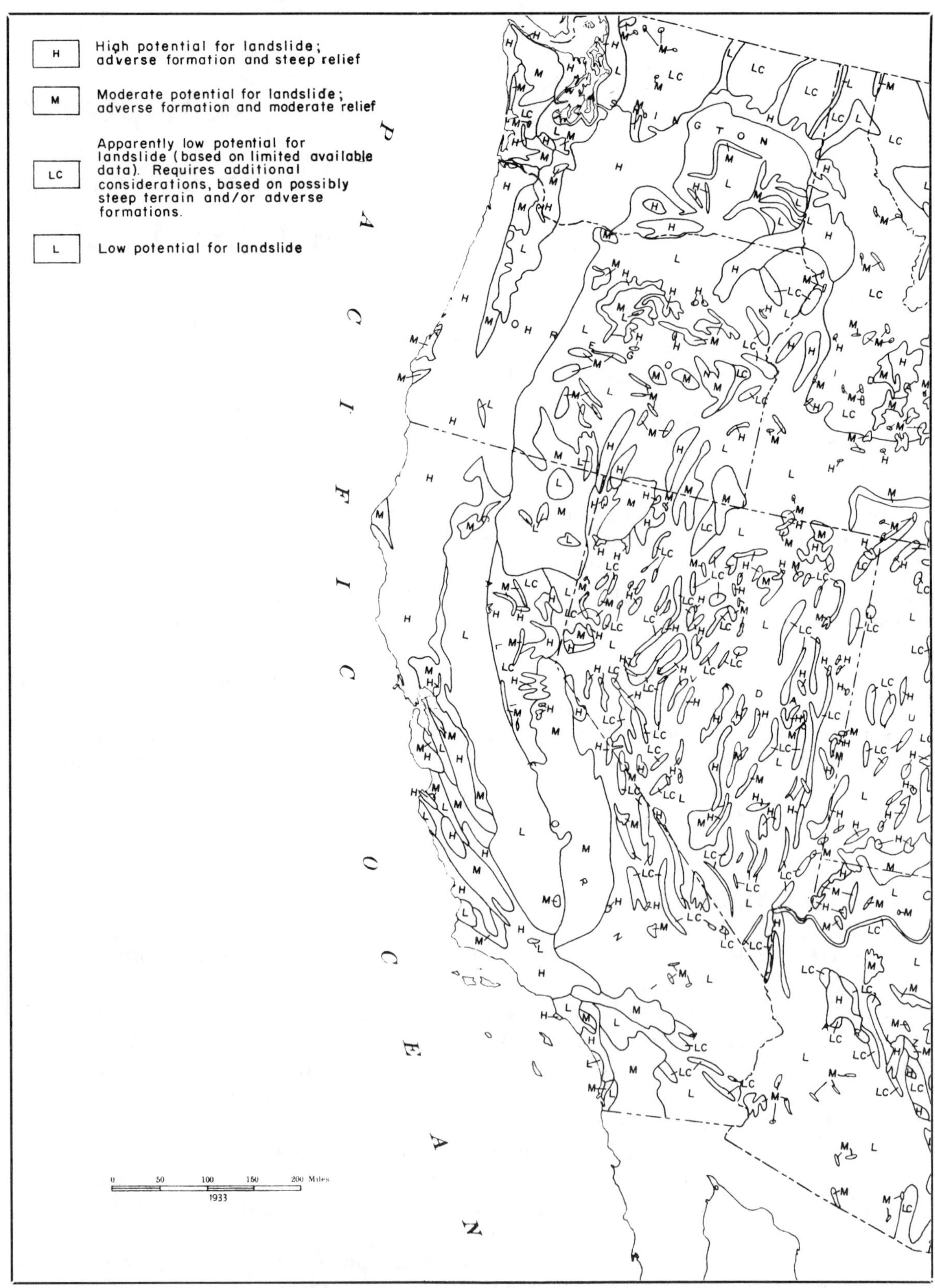

FIGURE 8-1. Interpretive landslide potential map. (From Krohn, J.P., and Slosson, J.E., "Landslide Potential in the United States," Calif. Geol., Oct. 1976, p. 224.)

TABLE 8-1
MASS WASTING

Type of Movement			Type of Material		
				Engineering Soils	
			Bedrock	Predominantly Coarse	Predominantly Fine
Falls			Rock	Debris fall	Earth fall
Topples			Rock topple	Debris topple	Earth topple
Slides	Rotational	Few units	Rock slump	Debris slump	Earth slump
	Translational	Few units	Rock block slide	Debris block slide	Earth block slide
		Many units	Rock slide	Debris slide	Earth slide
Lateral Spreads			Rock spread	Debris spread	Earth spread
Flows			Rock flow (deep creep)	Debris flow (soil creep)	Earth flow (soil creep)
Complex	Combination of two or more principal types of movement				

called a **block glide** or slide (Fig. 8-2), and failure along an arcuate or curved surface is called a **slump** (Fig. 8-3). Ranges of rates of movement are shown in Figure 8-4.

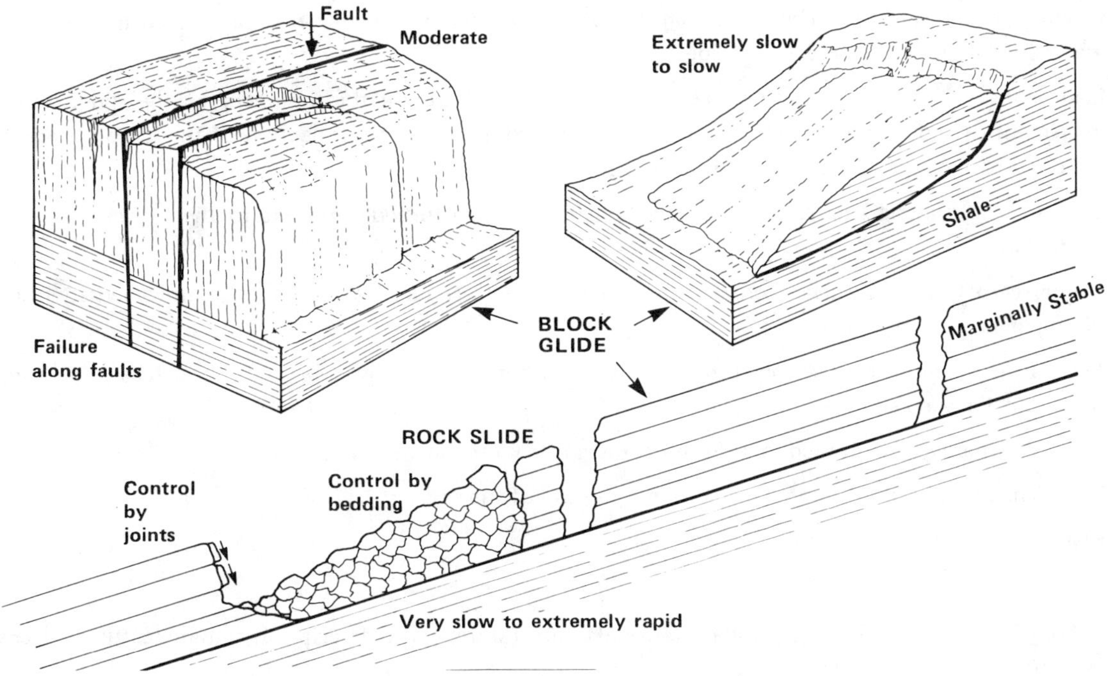

FIGURE 8-2. Examples of a block glide (slide). (Highway Research Board.)

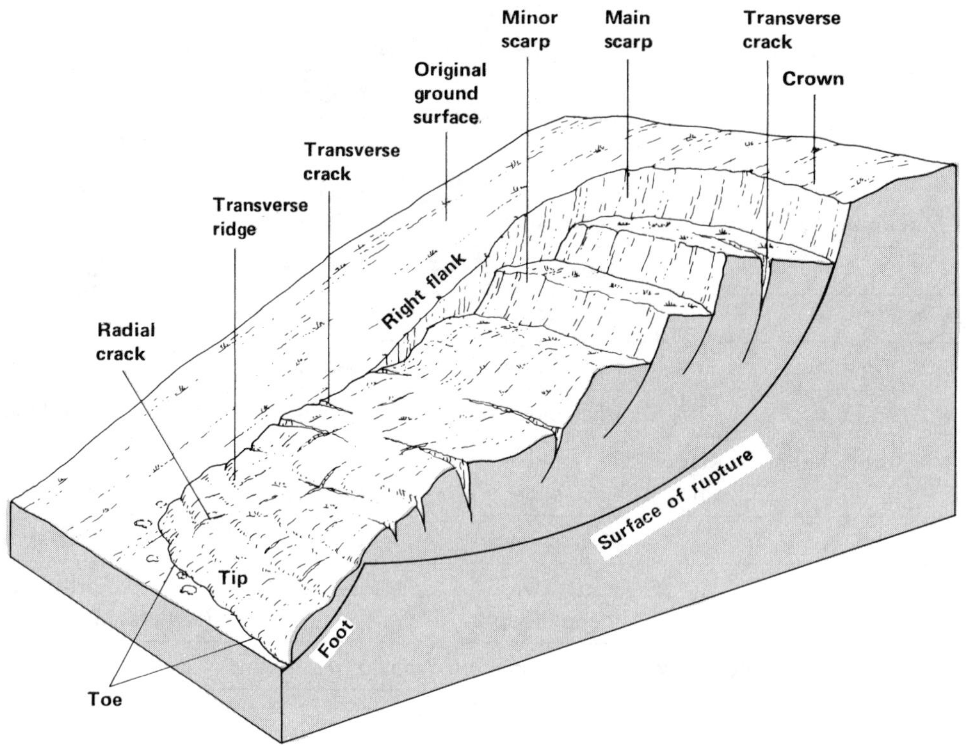

EXPLANATION OF TERMS

Original ground surface The slope that existed before the movement being considered took place. This may be the surface of an older landslide.

Main scarp A steep surface on the undisturbed ground around the periphery of the slide, caused by movement of slide material away from the undisturbed ground. The projection of the scarp surface under the disturbed material becomes the **surface of rupture.**

Crown The material that is still in place, practically undisturbed, adjacent to the highest parts of the main scarp.

Minor scarp A steep surface on the disturbed material produced by differential movements within the sliding mass.

Foot The line of intersection (sometimes buried) between the lower part of the surface of rupture and the original ground surface.

Toe The margin of disturbed material most distant from the main scarp.

Tip The point on the toe most distant from the top of the slide.

Flank The side of the landslide.

FIGURE 8-3a. Block diagram of a rotational failure (slump) with description of main features (Highway Research Board).

FIGURE 8-3b. Huge slump along highway near Orinda, California, 1951 (San Francisco Chronicle).

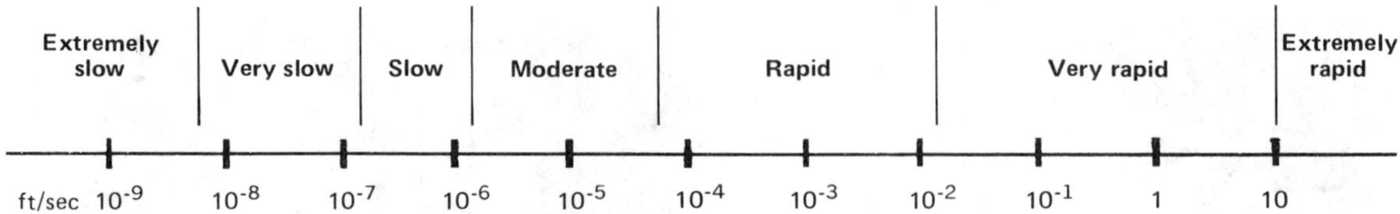

FIGURE 8-4. Classifying rates of landslide movement.

CAUSES OF LANDSLIDES

A slope fails when the driving forces, those tending to pull the slope down, exceed the resisting forces, those which support it and maintain stability. The driving force is gravity, whereas the resisting forces are largely frictional. An analogy is tilting a board upon which rests a piece of bread. The bread will remain stationary on the board until the tilt reaches a critical angle. At that angle the **driving forces** (D) equal or exceed the **resisting forces** (R) and the bread will begin to slide. The ratio of R:D is known as the **safety factor** (SF). When the SF is greater than unity a slide will not occur, when less than unity a slide will occur. It should be understood that a slope doesn't just fail; rather, changing conditions must lead to a decrease in resisting forces or an increase in the driving forces.

A number of factors, both natural and man-induced, may contribute to slope instability and landsliding. Here are the main ones.

1. **Oversteepened slopes.** When slopes are oversteep the resisting mass at the toe of the slope is lessened and potential failure planes, such as stratification, may be exposed.

2. **Loading the top of the slope.** Heavy loads, such as compacted fill, placed at the top of a slope increase the driving forces tremendously and may lead to failure. Curiously enough, a single-family dwelling does not add significant weight and does not enter into the calculation of slope stability. However, fills or heavy structures, such as warehouses or libraries with heavy dead loads, would be significant.

3. **Increased water content.** Excess water in the rocks or soils is a leading cause of landslides. Contrary to popular opinion, water does not act as a lubricant; rather, it increases pore water pressure, changes the effective stress between mineral grains, and adds weight to the slope. In this manner water serves to reduce resisting forces and adds to driving forces.

4. **Removal of support at the base of the slope.** Cutting away the toe of the slope is a common procedure to create more space for a building pad. In so doing the resisting forces are weakened and failure may occur if retaining structures are not built.

5. **Adverse bedding or rock structures.** Exposed (daylighted) bedding or joint planes offer ready-made planar surfaces upon which rock masses may slide (Fig. 8-5). These planes may be exposed by careless or thoughtless manufacture of cut slopes for highways or subdivisions (see 1 above).

6. **Triggering action.** Repeated vibratory stresses such as earthquake shaking, heavy traffic, or sonic booms may cause slopes to fail that are marginally stable.

PREVENTION AND CORRECTION OF LANDSLIDES

Although landslides are nature's way of achieving slope stability and topographic equilibrium, they represent a significant geologic hazard. It is projected that, in California alone, landslides will result in $10 billion damage by the year 2000 (Alfors, J. T., and others, 1973, Urban Geology, Master Plan for California, Calif. Div. Mines and Geology Bulletin 198). However, 90 percent of this estimated loss may be prevented by avoiding slide-prone areas and by applying all possible preventative and remedial measures. These include preliminary site investigations by geologists and soils engineers and appropriate grading ordinances by local government. The costs of preventative measures are estimated to be 10 percent of the projected loss, thus we have a favorable cost: benefit ratio. Methods of control and prevention are fairly straightforward and are relatively simple if recognized early in the planning process.

Drainage Control. Inasmuch as surface and subsurface water are the major cause of landslides, controlling them is usually the most effective means of preventing problems. The general strategy is to prevent surface-water runoff from creating gullies and causing excessive slope erosion, and to prevent water infiltration, which, if unchecked, builds up seepage forces and weakens slopes. There are various methods for controlling building site drainage.

(a) Proper building-pad slope that directs water to adjacent streets and acceptable storm drainage systems. Local building codes usually require a slope of 2 percent (2-ft drop per 100 ft horizontally) from back to front of the lot. Concrete gutters are sometimes used to collect roof drainage and convey it offsite.

(b) Planting deep-rooted vegetation, such as Algerian Ivy, to prevent soil erosion.

(c) Covering the slope with a sealant or large plastic sheets if soil erosion has already begun.

(d) Creating a surface of compacted impermeable soil 18–24 in. thick, to prevent seepage of water into exposed bedding or joint planes (Fig. 8-6a). This fill is known as a **blanket seal** and is prescribed in local building codes where important.

(e) Preventing subsurface water from accumulating by drilling horizontal holes and installing perforated plastic pipe, usually 2–3 in. in diameter. These drains are known as **hydrauger holes** and are quite efficient if properly designed (Fig. 8-6a, b).

Proper Grading. This involves reducing the steepness of the slope or removing material from the top of the slope and placing it as compacted fill at the toe. Safety of manufactured slopes

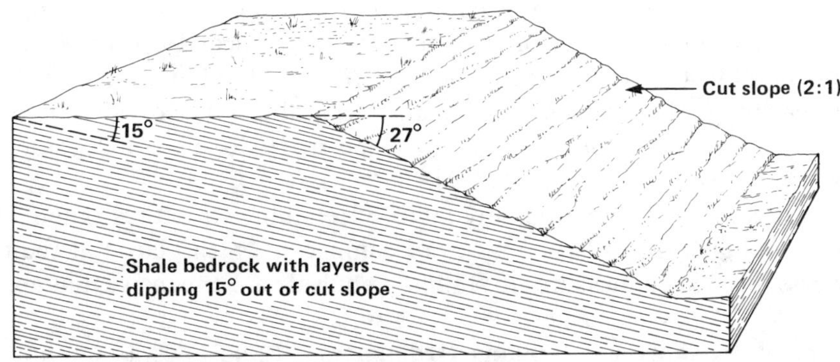

FIGURE 8-5. Diagram illustrating "daylighted" bedding out of a cut slope. This condition leads to block glides along lowest layer exposed.

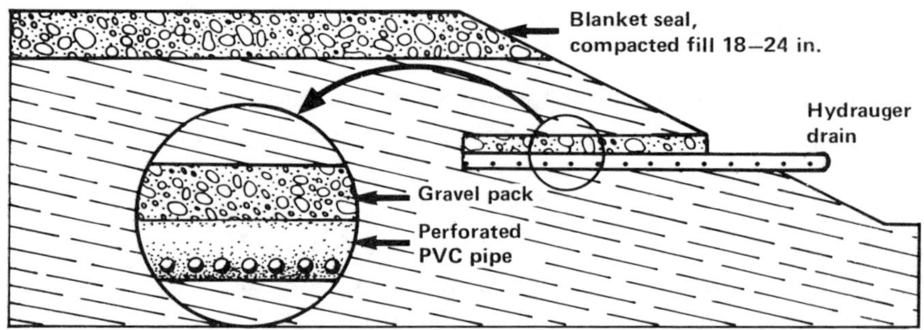

FIGURE 8-6a. Diagram illustrating blanket seals and hydrauger drains. Both these devices prevent water from accumulating in the earth materials supporting the slope.

FIGURE 8-6b. Hydrauger drains and header-pipe collector installed into a slope in San Leandro, California.

is usually calculated by a soils engineer in conjunction with a geologist. The slope steepness is usually expressed as a ratio of the horizontal distance to the vertical drop. Thus a 2:1 slope has a slope of 2 ft horizontal to 1 ft vertical (equivalent to 27° from the horizontal). This may also be expressed as a percentage slope which is defined as vertical distance divided by the horizontal distance times 100. Thus a 2:1 slope is also a 50% slope, a 1:1 slope is a 100% slope, and a 1.4:1 slope is equivalent to a slope of 70% (Fig. 8-7).

Retaining Structures. There are many kinds of retaining structures ranging from simple concrete block retaining walls to large masses of compacted earth placed at the toe of slopes known as "buttress fills." Such measures to stabilize a hillside are expensive, but are appropriate in certain cases.

Wise Planning Decisions and Grading Ordinances. Landslides are common in hillside areas and have caused extensive damage to private and public property. Until we begin to appreciate the causes of slope instability and plan our engineering works with these in mind, landslides will continue to be a major geologic hazard. A detailed preliminary investigation should be conducted by qualified professionals before any major construction in an area of significant relief. This investigation should describe existing slides in the area and delineate limited use areas and areas subject to other geologic hazards, such as inundation and fault rupture.

RECOGNIZING LANDSLIDE TERRAIN

To prevent future tragedy we need to recognize old landslides and landslide terrain. Areas of historic landslides usually exhibit anomalous topography, that is, numerous depressions and scarps, inclined trees, anomalous flat areas, and distinct vegetation differences. In Figure 8-3 we can see that after extensive erosion the slump illustrated would show rounded hills and swales that characterize hummocky topography. Figure 8-8 is a reproduction of part of the Pacific Palisades, California, and was one of the first attempts to map landslides in an urban area. Note the large slide marked A (for active) that has moved onto the Pacific Coast Highway, a major interstate highway. Uniform slopes change from steep along the scarp marking the top of the slide to gentle topographic benches further downslope. We recognize slides on maps from topographic benches and from contours that bulge outward downslope. Usually erosion creates gullies on either side of the slide curving upward and inward toward the crown scarp.

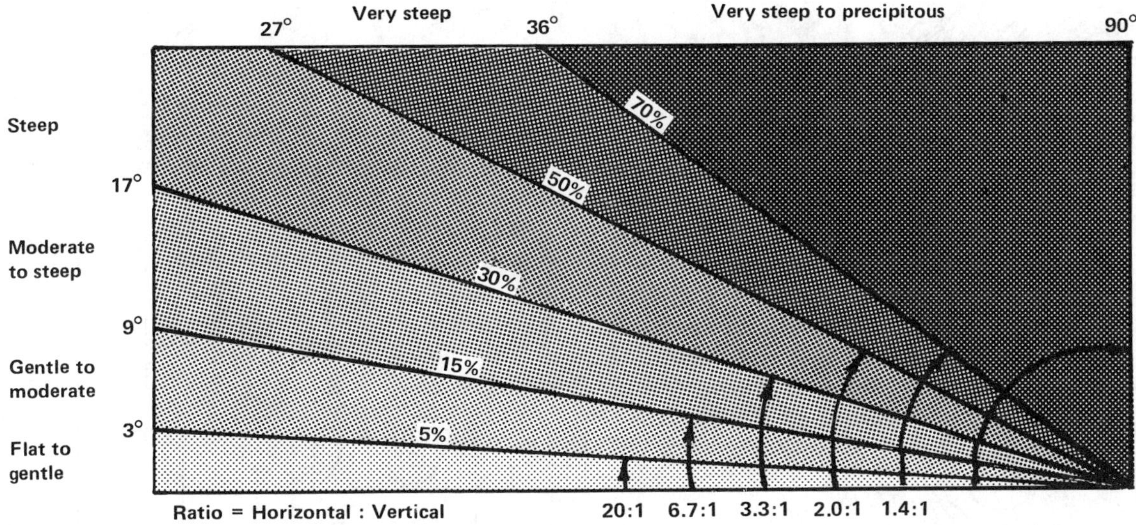

FIGURE 8-7. Categories of land slope expressed as percentages, degrees, and ratios.

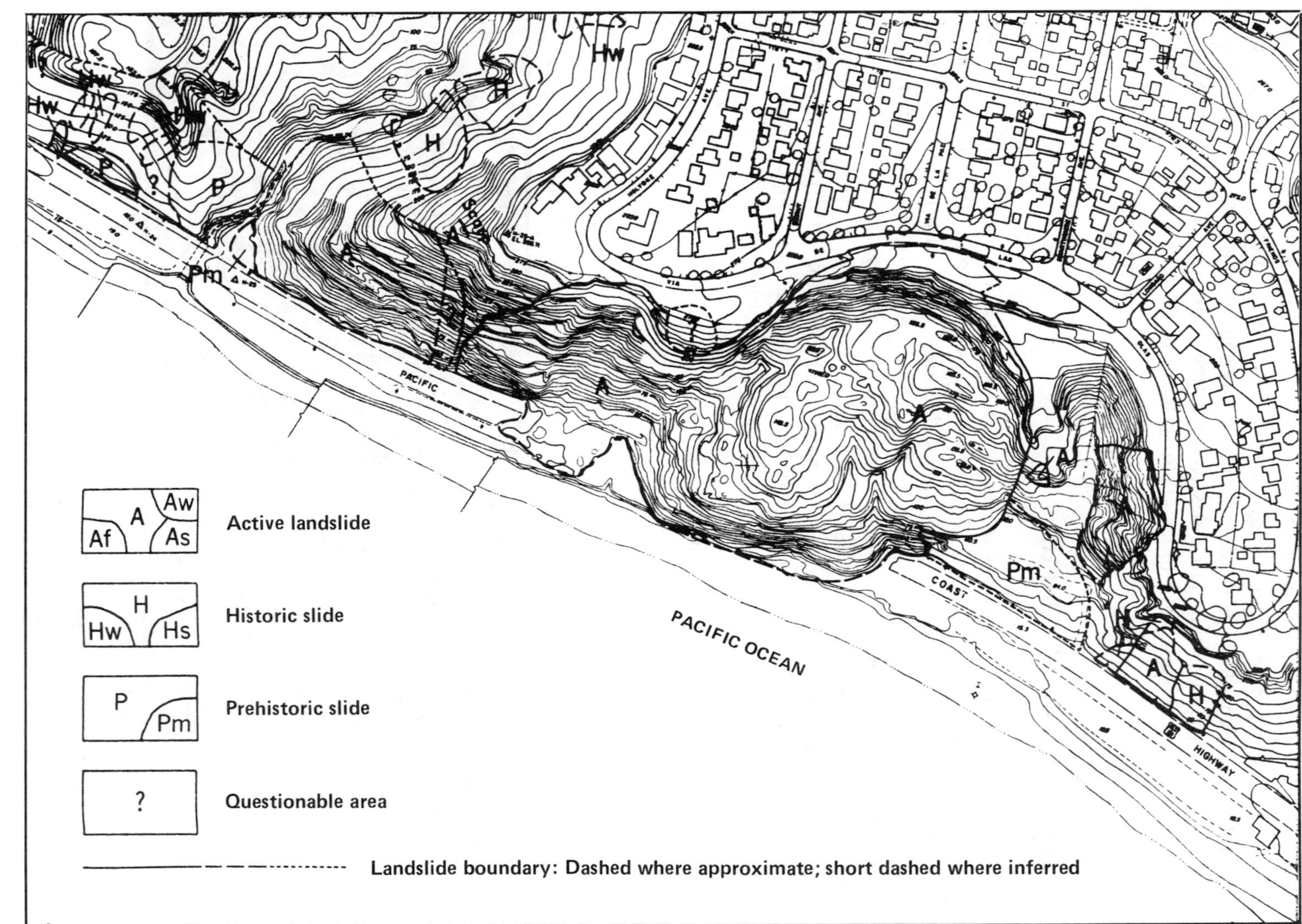

FIGURE 8-8. Map of huge landslide at Pacific Palisades, California. The coast highway was reconstructed around the toe of the slide Scale 1:4800. Contour interval 5 ft. (From McGill, J.T., 1959, U.S. Geol. Survey Misc. Geological Investigations Map I-284.)

EXERCISE VIII

LANDSLIDES AND MASS WASTING

Name ______________________________

1. The cross-sections in Figure 8-9 represent slopes exposing different geologic conditions. For each cross-section give the information requested.*

   (a) What is the stability of the slope shown? If you conclude it is unstable or metastable please describe the geology leading to this condition.

   ____

   ____

   ____

   ____

   (b) Sketch on the cross-section how the slope might fail.

   (c) Illustrate and describe how the slope might be modified to eliminate or reduce the hazard.

2. Study the part of the Clayton quadrangle, California, shown in Figure 8-10. Many landslides exist that can be recognized by anomalous topography and areas of highly irregular and jagged contour lines. One large slide area covering about half a square mile is shown in the northeast part of the map.

   (a) Outline as many slides on the map as you can find that fit the critieria given. Note that the map symbol for a landslide consists of outlining the boundary of the slide and showing the direction of motion with single-barbed arrows.

---

*The authors thank the Council on Education in the Geological Sciences (CEGS) for suggestions and ideas contained in their CEGS Publication 15: Pestrong, Raymond, 1974, Slope Stability; New York: McGraw-Hill Book Co.

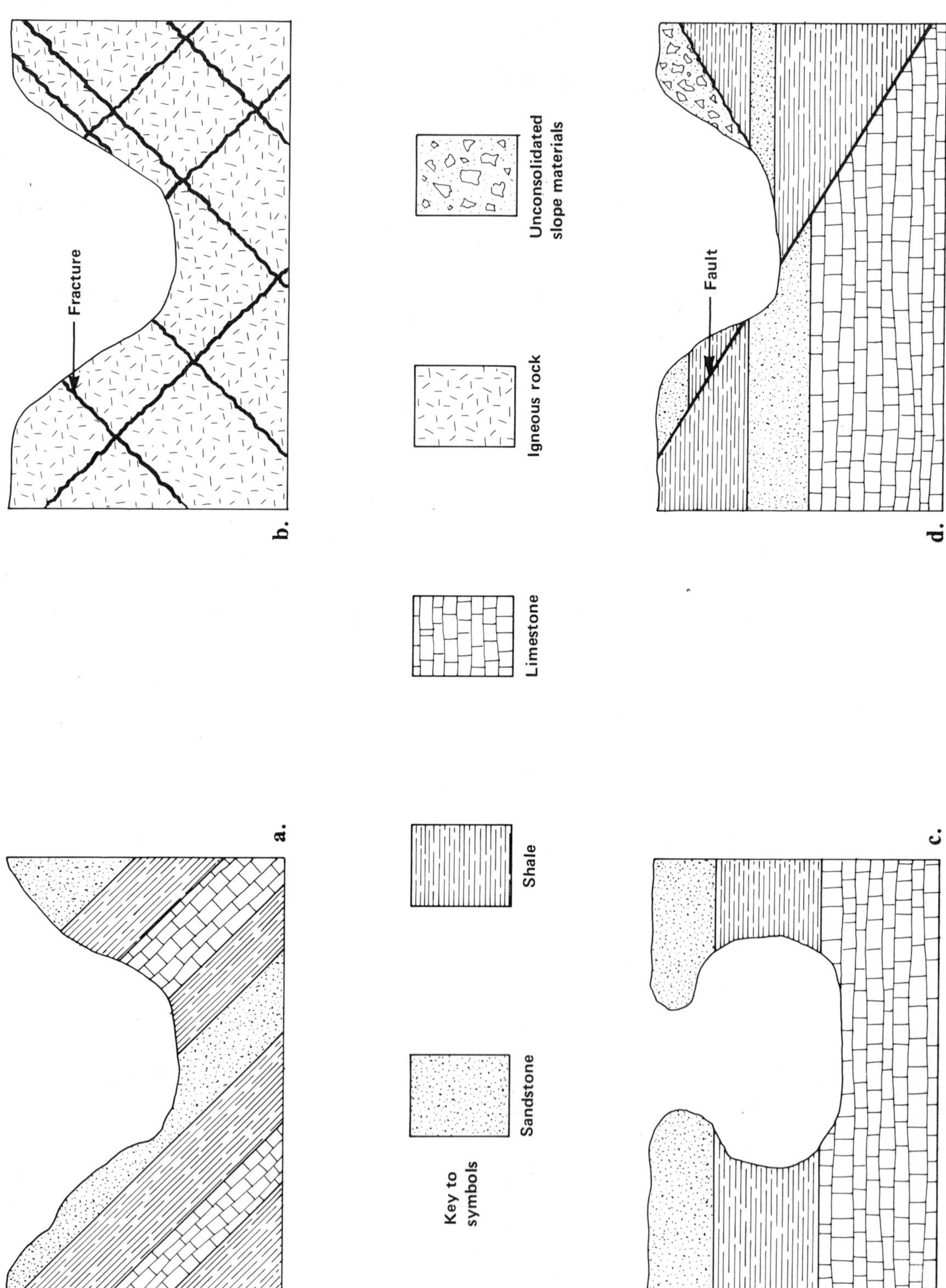

FIGURE 8-9. Cross-sections of valleys exposing different geologic conditions.

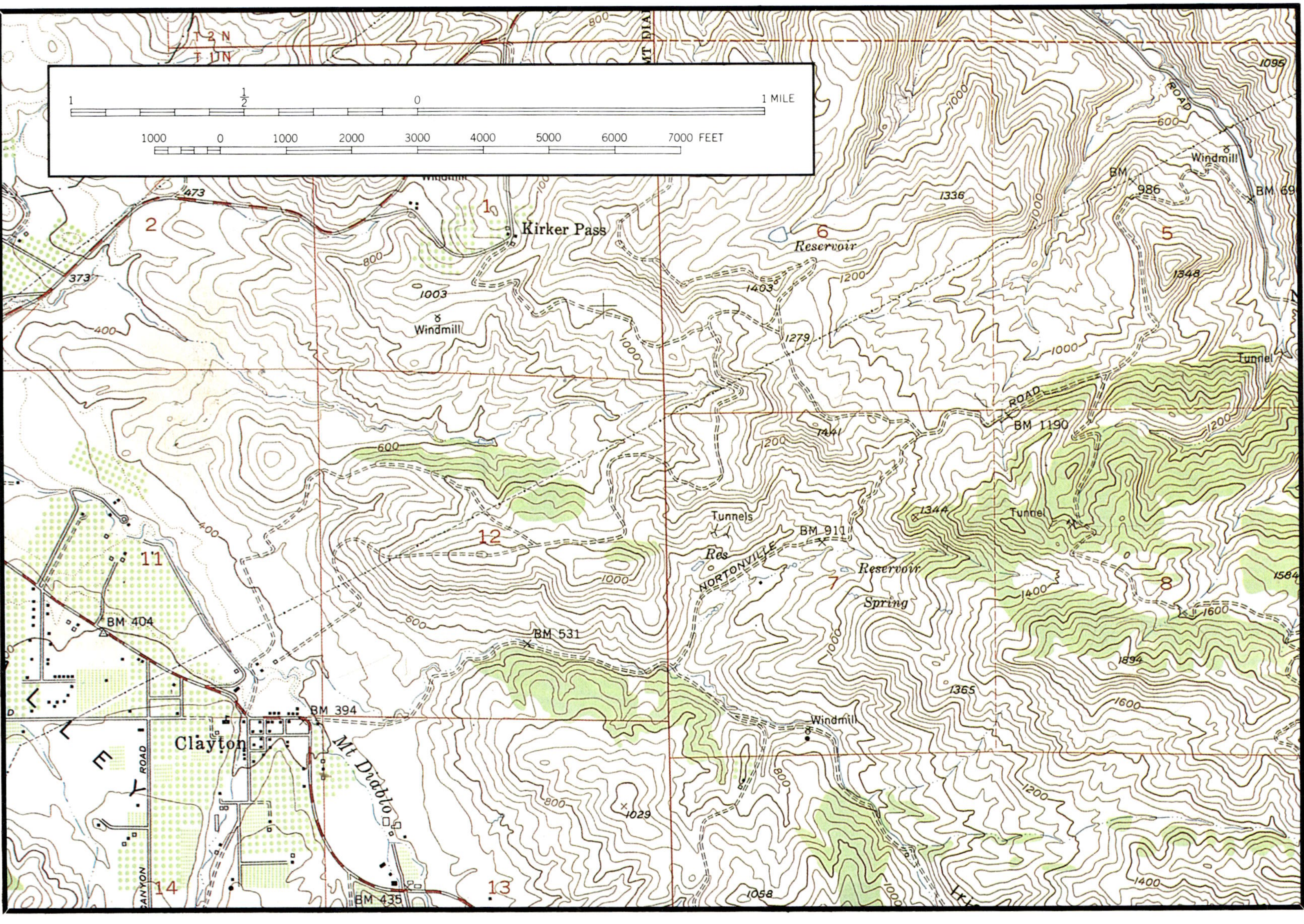

FIGURE 8-10. Portion of Clayton quadrangle, California. Contour interval 40 ft; dotted contour = ½-interval. (Courtesy U.S. Geol. Survey.)

(b) The east-west trending ridge in section 12 has fewer slides than the similar trending ridges to the north and south. Give a reasonable geologic explanation for this fact.

____

____

____

(c) Choose a site for your home that overlooks the town of Clayton and mark it on the map. Remember, you need access to the home site and support facilities (sewage disposal, water, etc.).

3. Make simple line sketches illustrating the following slopes.

2:1

1:1

1½:1

4. As a geologic consultant for a potential buyer you are asked to evaluate a hillside region in terms of its suitability for a housing development. What features or conditions would you investigate and what data would you require to make an intelligent decision? How may this data be obtained?

____

____

Chapter 9
Surface Water

The waters of the Earth can conveniently be divided into (1) surface water and (2) subsurface water. Surface water includes rivers, lakes, and oceans; subsurface water includes ground water, connate water, and juvenile water. In this chapter, we will focus our attention on surface water.

THE HYDROLOGIC CYCLE

The **hydrologic cycle** describes the movement of water at and near the surface of the Earth (Fig. 9-1). Water moves from the Earth's surface into the atmosphere by **evaporation** of surface water and water in the soil near the surface, and by **evapotranspiration** of water from

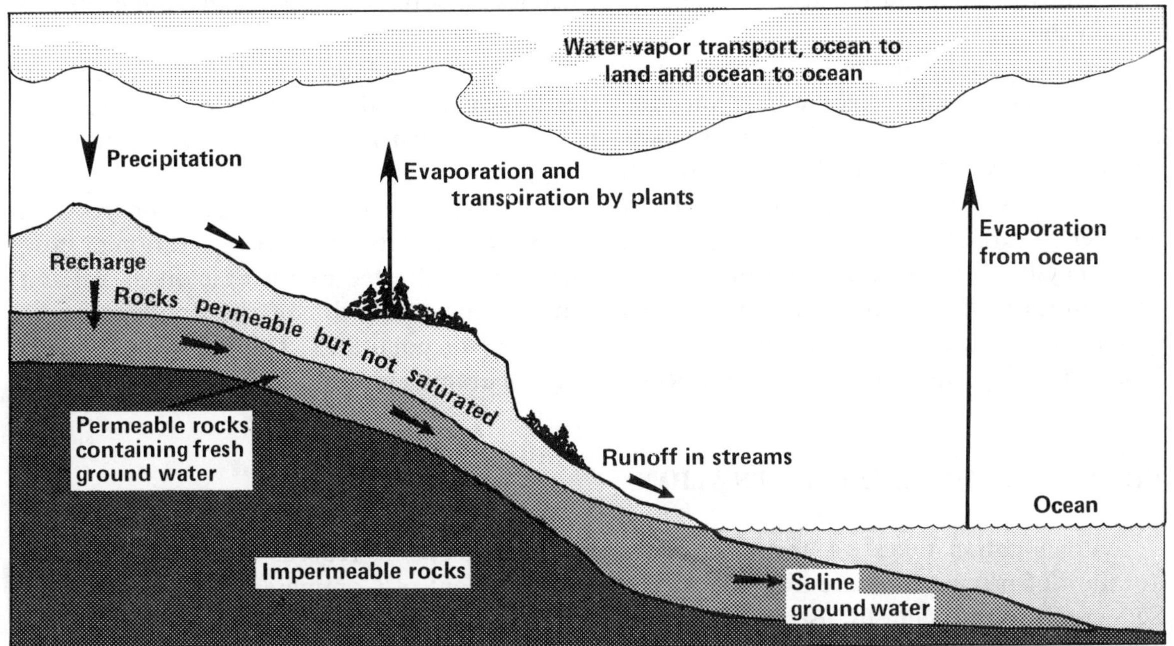

FIGURE 9-1. The hydrologic cycle (U.S. Geol. Survey Circular 601-I).

plants. The evaporated water is in the form of vapor. The atmospheric water vapor is moved by air masses and eventually condenses to form clouds. The clouds may result in some form of precipitation, such as rain, snow, or hail, over the oceans or over the land surface. The water that falls on the land surface moves (1) on the surface as sheet flow or in streams and rivers, and (2) through the soil into plants or into the ground water. With the aid of gravity, much of the surface water and ground water eventually flows into the oceans, although, because of evaporation and evapotranspiration, a large amount does not. Thus, the hydrologic cycle provides a general picture of the movement of water and serves to focus attention on several concerns of environmental geology: movement of surface water, including flood waters; storage of surface water in reservoirs, lakes, and oceans; movement and storage of ground water; evaporation and evapotranspiration; and precipitation.

The hydrologic cycle implies a uniform exchange of water from the surface to the atmosphere and back to the surface. This may be valid worldwide, but locally the cycle is not uniform. Such non-uniformity in the cycle can result in locally excess precipitation and flood or locally deficient preciptation and drought. The hydrologic cycle and the local variations provide the backdrop against which we can accommodate the effects of the variations. We can, for example, mitigate the effects of flood waters by building flood-control channels, or store water in reservoirs, or manage withdrawal of ground water.

ANALYSIS OF ENVIRONMENTAL HAZARDS

Rainfall that reaches the surface of the Earth may either flow along it or infiltrate the ground. In rural areas, surface water flows into gullies, creeks, and rivers. In urban areas, it flows into gutters, sewers, and lined channels. Physically, rural and urban areas are different and present different problems. However, the principles for solving the environmental problems remain the same.

The environmental or engineering geologist is concerned with both the beneficial and non-beneficial aspects of surface water. Surface water is beneficial when it supplies our needs for consumption, recreation, irrigation, or industry. It is not beneficial, or at least a nuisance, when it results in floods or becomes polluted.

Analysis of the environmental problems or hazards of surface water usually focuses on a number of variables: amount of water flowing in a stream and its tributaries, duration of flow, frequency of flooding, and water quality. These are in turn affected by such factors as amount and duration of precipitation, extent of infiltration, shape and size of the drainage basin, amount and velocity of stream water, and the presence of pollutants. The following sections briefly describe some of the most significant of these parameters.

PRECIPITATION AND INFILTRATION

Precipitation takes the form of rain, snow, or hail. While records of the amount and duration of precipitation are available from the Department of Commerce, National Weather Service, it is difficult to use precipitation data directly for solving many environmental geologic problems because of the complexities in determining the total rainfall in an area, the amounts of evaporation and evapotranspiration, and the amount of infiltration. To a great extent, these difficulties are resolved by using stream flow measurements, which are velocity and amount of water in a stream. Such measurements are simple to make and simple to use.

The amount of water that infiltrates the ground is a function of the rock and soil types and their **permeability**. Permeability is how well the rock or soil allows water to flow through it. This ability of the water to flow through depends on the volume of voids or pore openings between the grains and on how the pore openings are connected: the more connections of pore openings, the greater the permeability. **Porosity** is the ratio of the volume of the pore openings to the total volume of the rock or soils. A rock may be porous and not permeable. (How could such a condition occur? Name a rock that is porous but not permeable.) If the rocks are permeable, then water can infiltrate relatively easily. The greater the capacity for infiltration, the lower will be the amount of water at the surface for runoff. In areas where the soil is thin and the bedrock is impermeable, surface runoff can be rapid, which in turn may allow the rivers to reach flood stage relatively quickly.

DRAINAGE BASIN

The **drainage basin** or watershed of a stream encompasses the area that collects and directs the surface water into the stream. Drainage basins are separated by a **drainage divide**, which is defined by the topographic high between the basins (Figure 9-2). A basin may be just

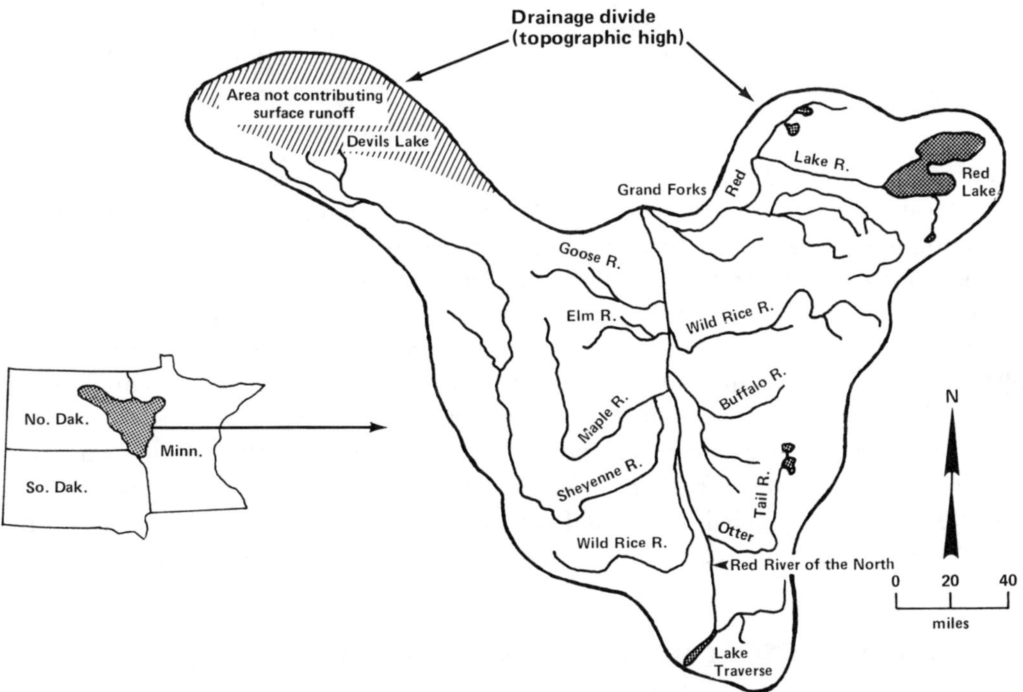

FIGURE 9-2. Drainage basin of the Red River (Harrison, S. S., 1968, "The Flood Problem in Grand Forks–East Grand Forks," North Dakota Geol. Survey Misc. Series 35).

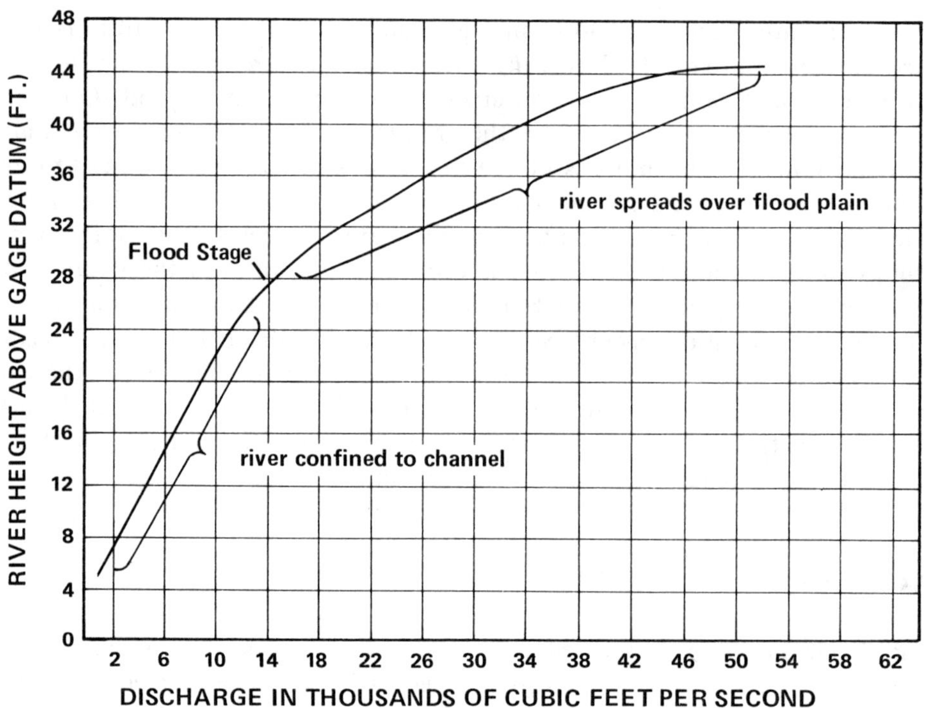

FIGURE 9-3. Discharge-river height curve for the Red River at Grand Forks showing relationship of river height to stream discharge. This kind of plot is called a rating curve and gives information of flood stages of the river in question (Harrison, S. S., 1968, "The Flood Problem in Grand Forks–East Grand Forks," North Dakota Geol. Survey Misc. Series 35).

the area on the sides and head of a gully, or it may be a complex system of rills, creeks, and streams that drain into a major river. The size of a drainage basin can be determined directly from a topographic map or an aerial photograph simply by drawing a line along the drainage divide and measuring the area within that line. The direction of surface water flow tends to be perpendicular to the countour lines.

The shape of a drainage basin can affect the rates of the flow of surface runoff. For example, if the water runs in long and narrow tributary channels, the flow would be relatively slow and would arrive at the main river at different times depending on the length of the tributary; consequently, the river may be able to accommodate the inflow over a relatively long period of time. However, if the water runs in wide and short tributary channels, the flow would be relatively rapid and would arrive at the main river in a relatively short period of time. If the river cannot accommodate the rapid inflow, flooding results.

VELOCITY AND AMOUNT OF WATER IN A STREAM

The flow of water in a stream is not constant. The variation results from changes in the amount of water entering the stream, which in turn depends on the amount of precipitation. The velocity of water depends on the slope of the stream bed, its shape, and the amount of water in the stream. However, if we keep the slope and shape constant, then the velocity of the stream flow increases with an increasing amount of water.

The stream flow is measured at various points along the stream by determining the cross-sectional area and velocity of the water. The cross-sectional area (A) is found by measuring (1) the bottom topography of the stream at a station and (2) the height of the water above an arbitrary datum plane. The height of the water is called the **river gage** or **stage**. The velocity (V) of the stream is measured by using a current meter. The amount of water flowing through the cross-sectional area may be expressed as $Q = AV$, where Q (called discharge) is usually given in cubic feet per second. One cubic foot contains 7.48 gallons.

The discharge may be plotted against the gage station height to develop a **rating curve** for that station. A typical rating curve is shown in Figure 9-3. The rating curve thus established can be used to determine the discharge for other gage heights by measuring the gage heights in the field and comparing their values to the graph. More importantly, the rating curve also supplies information that can be used to control floods. According to the discharge-river height curve for the Red River at Grand Forks, North Dakota, a discharge of 3,000 cubic feet per second (cfs) is needed to raise the river level from 20 to 25 ft above the datum plane. However, it requires 4,500 cfs to raise the river from 25 to 30 ft above the datum plane. The reason for this sharp increase is that at 28 ft, the river overflows and greatly widens its channel. For the same reason, the curve is much flatter above 28 ft than below it; that is, it requires a greater discharge during overflow conditions to raise the river level one foot than under normal flow regimes. Flood and river-valley terminology are illustrated in Figure 9-4.

Hydrograph. A hydrograph is a plot of discharge or stage (water height) against time and shows the nature of the rise and fall of a river for a given discharge period. The graph can be used to determine the total flow over a period of time, periods of greater or lesser flow, and

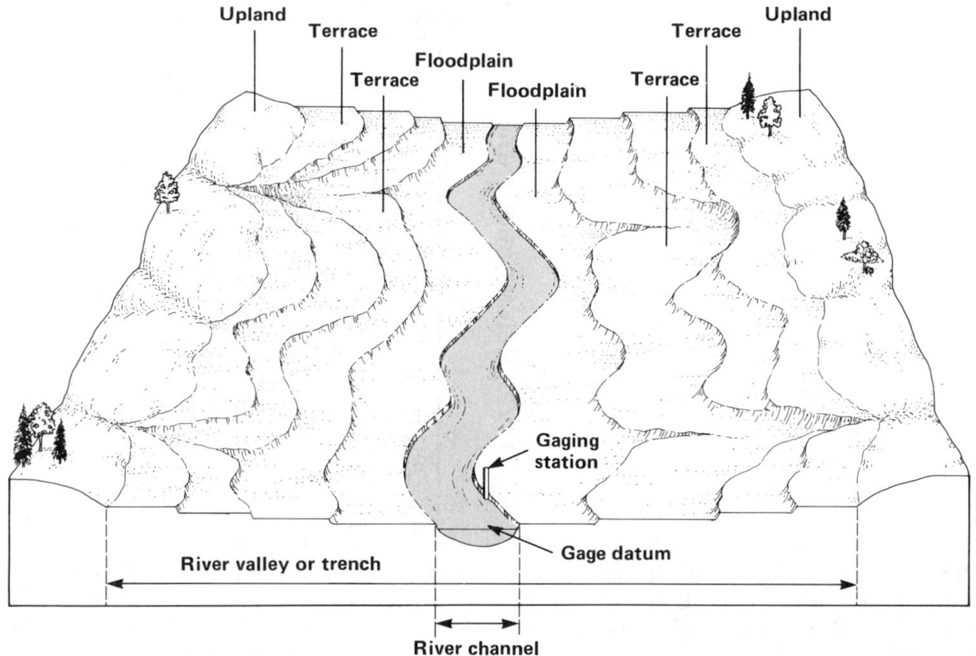

FIGURE 9-4. River channel and flood-plain terminology.

peak flow or discharge. Such information is invaluable in flood-control studies. Two such hydrographs are shown in Figure 9-5, one for the flood of 1897 on the Red River in North Dakota (a) and the other for an arid region subject to flash flooding (b). Note the difference in duration of water-level fluctuation between the two graphs: hours for the Tucson flood and weeks for the Red River. This clearly reflects climatic and topographic differences and illustrates the "flash flood" of about 8 hours duration typical in the arid southwest and the strong seasonal flood involving rapid snow melt and rain common in the north central United States.

Flood Frequency. One of the most useful relationships that can be derived from flood records is that of flood frequency or recurrence interval. With this information we are better

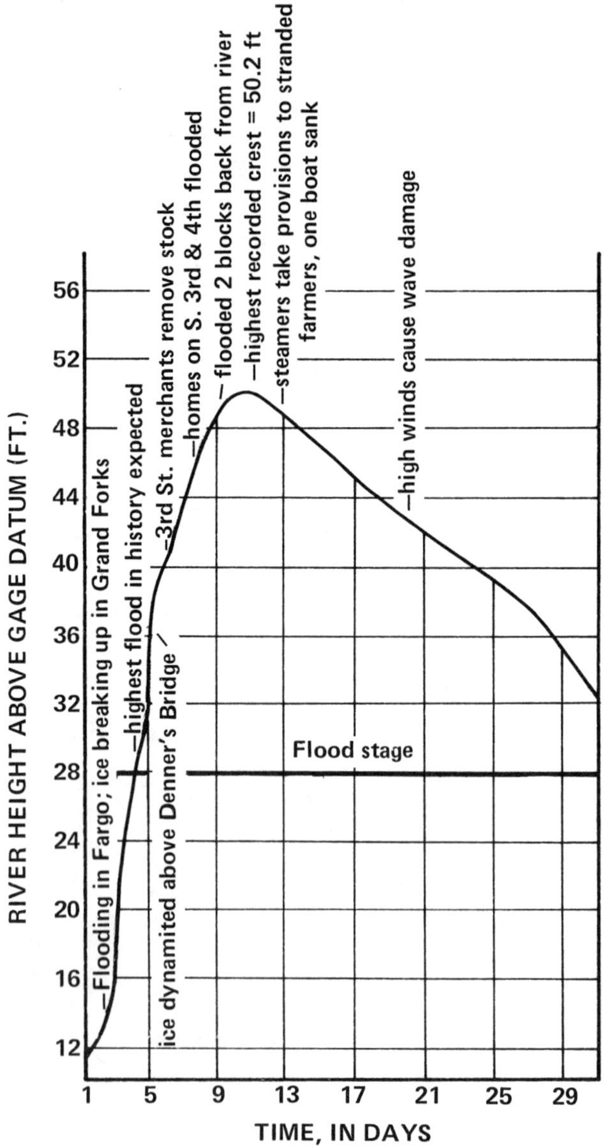

FIGURE 9-5a. Stage hydrograph for the great flood of 1897 on the Red River in North Dakota (Harrison, S. S., 1968, "The Flood Problem in Grand Forks–East Grand Forks," North Dakota Geol. Survey Misc. Series 35).

equipped to design dams, bridge clearances, size of storm drains, and the like. The data required are the values of maximum or peak discharge that occurred each year over a long period of time. The longer the time period the more reliable is the resulting statistical analysis.

The peak discharges are tabulated and ranked according to their magnitudes—the highest discharge being ranked as "1," the second maximum discharge ranked as "2," and so forth. The recurrence interval or return period can be computed using the equation

$$T = \frac{N + 1}{M}$$

where T is the recurrence interval, N is the number of years of observation, and M is the magnitude of the maximum discharge. When maximum discharge is plotted against the corresponding recurrence interval we obtain a **flood-frequency graph** (Fig. 9-6). From the graph, we can estimate how often a flood or discharge of a particular magnitude should occur. For example, suppose we wish to build a warehouse with a useful life of 40 years on the Red River at Grand Forks. Figure 9-6 shows that the river reaches an elevation of 48 ft above the datum plane once every 40 years (the "forty year" flood). Therefore, we should find a site higher than 48 ft or add artificial fill on lower ground to raise the base of the structure above that level.

Another way of looking at recurrence intervals is in terms of the likelihood or probability of a given flood level occurring in any one year. In our example you might ask, "What is the

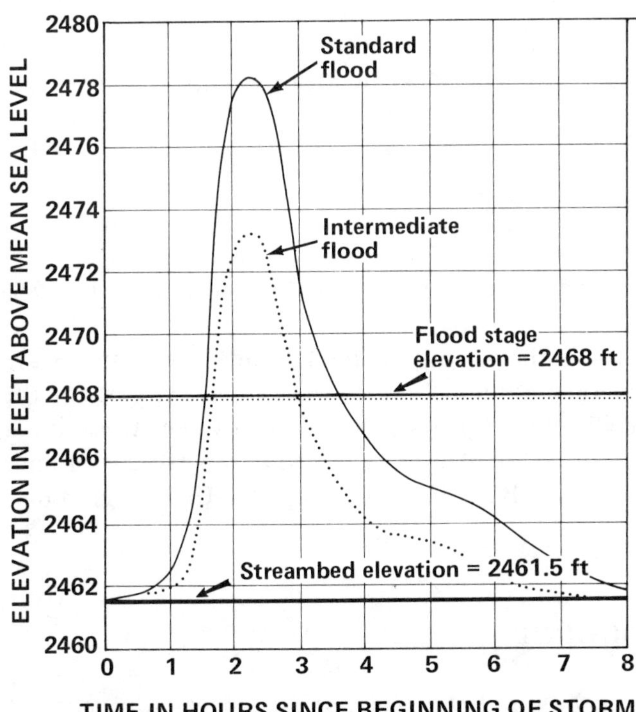

FIGURE 9-5b. Stage hydrograph for a 5-hour storm, August 1975, on the Tanque Verde Creek in Tucson, Arizona. Notice the marked difference between North Dakota and Arizona in duration of flooding.

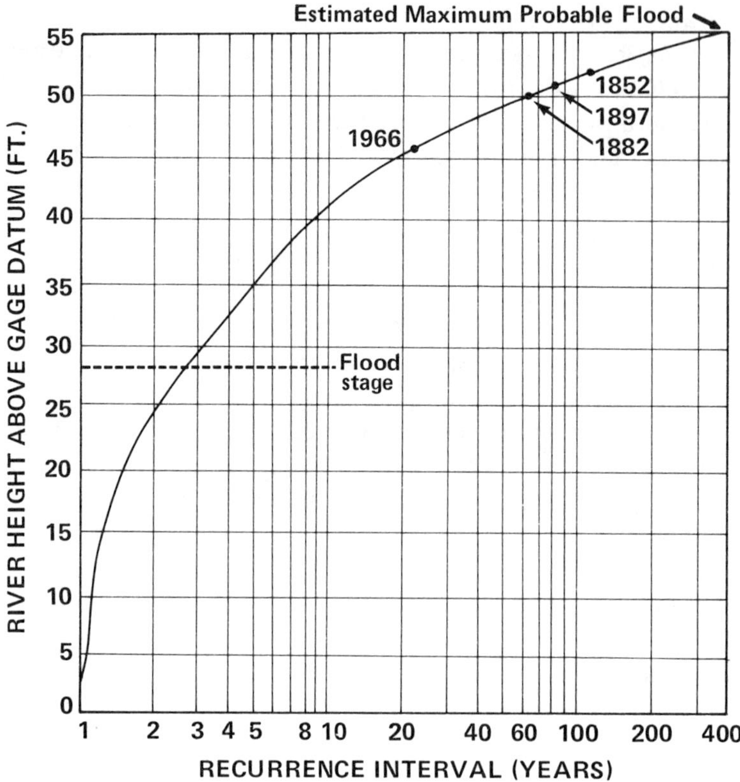

FIGURE 9-6. Flood-frequency graph for the Red River at Grand Forks, North Dakota.

chance of the 50-year flood occurring in any one year?" Such a probability is the inverse of recurrence interval or probability:

$$\frac{1}{T} = \frac{1}{50} = .02 \text{ or 2 percent}$$

In other words, there is a 2% chance in any one year for a 50-year flood to occur, or a 98% chance that such a flood won't happen.

The arithmetic mean of the peak flows is the **average annual flood**. The statistics are such that the recurrence interval of the average annual flood is the same regardless of the length of record. In our example it is specifically 2.3 years, which is equivalent to saying a flood of average magnitude can be expected to be equaled or exceeded on an average of once in 2.3 years, or 10 times in 23 years (Leopold, Luna B., 1968, "Hydrology for Urban Planning," U.S. Geol. Survey Circular 554).

EFFECT OF URBANIZATION ON FLOODING

To the geologist, urbanization means paving and this has a serious negative impact on the surface-water hydrology of a region. Specifically, urbanization increases total runoff and changes the peak-flow characteristics of the drainage basin or stream system in question. Two influential factors are (1) the percentage of the area sealed off to infiltration and (2) the rate at which runoff moves across the land and into stream channels. In residential areas of small lot size as much as 80% of the surface area may be rendered impermeable to water. This means that more rain and a greater extent of paved areas will increase rates of flow into storm drains and rivers. Naturally, the quality of the water is diminished in its travel through the urban environment.

Figure 9-7 is instructive and shows the effect of urbanization on a hypothetical hydrograph. Before urbanization there is a greater lag time between intense rainfall and peak stream flow. The total flow is distributed over a considerable length of time. After urbanization, lag time is shortened, peak flow is greatly increased, and total runoff is compressed into a shorter time interval. Remember, water runs off faster from roofs and paving than from vegetated areas, natural or man-made. In a city that is totally served by storm drains and where 60% of the land surface is impervious, floods are almost six times more numerous than before urbanization. Figure 9-8 shows increase in flooding (overflows) as a measure of urbanization.

Of course there are effects of urbanization other than those discussed. The most important are changes in sediment yield and water quality. Sediment yield may be more or less, depending on the stage of construction in the urban area. More exposed ground means more sediment yield and vice-versa. Water quality is greatly reduced by the introduction of sewage into streams and rivers and the inevitable collections of beer cans, garden cuttings, and trash that are dumped irresponsibly into local stream channels.

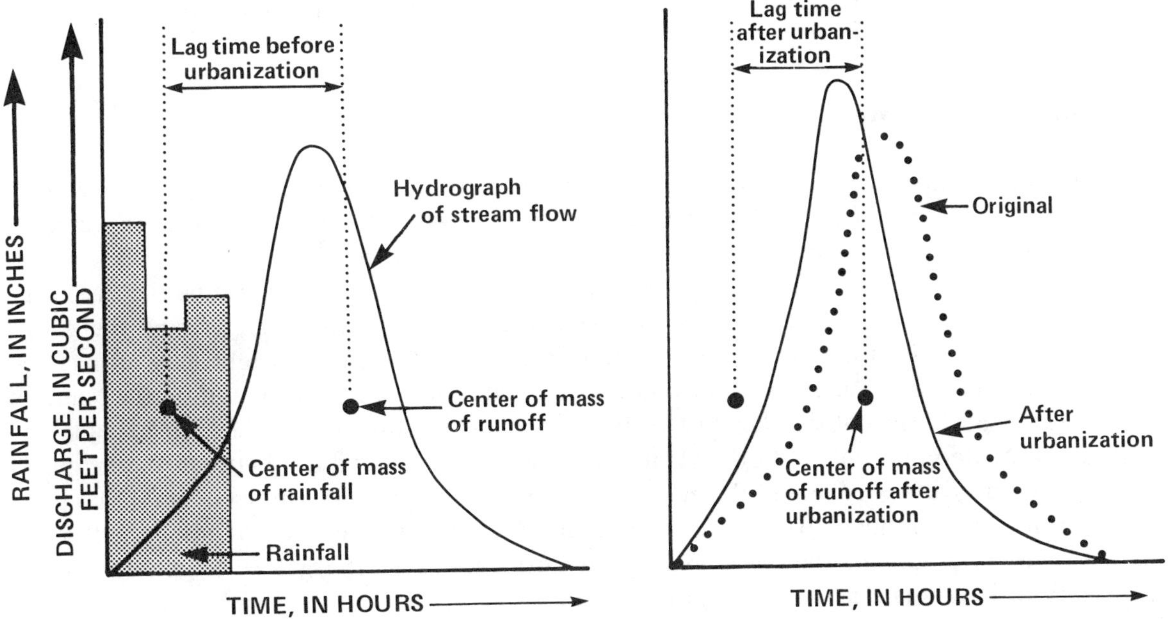

FIGURE 9-7. Unit hydrographs showing effects of urbanization on peak flow, duration, and lag time between intense rainfall and flooding. Note that the center of mass of the hydrograph of runoff after urbanization has shifted to the left (shorter time) compared with before urbanization. (Leopold, Luna B., 1968, "Hydrology for Urban Planning," U.S. Geol. Survey Circular 554.)

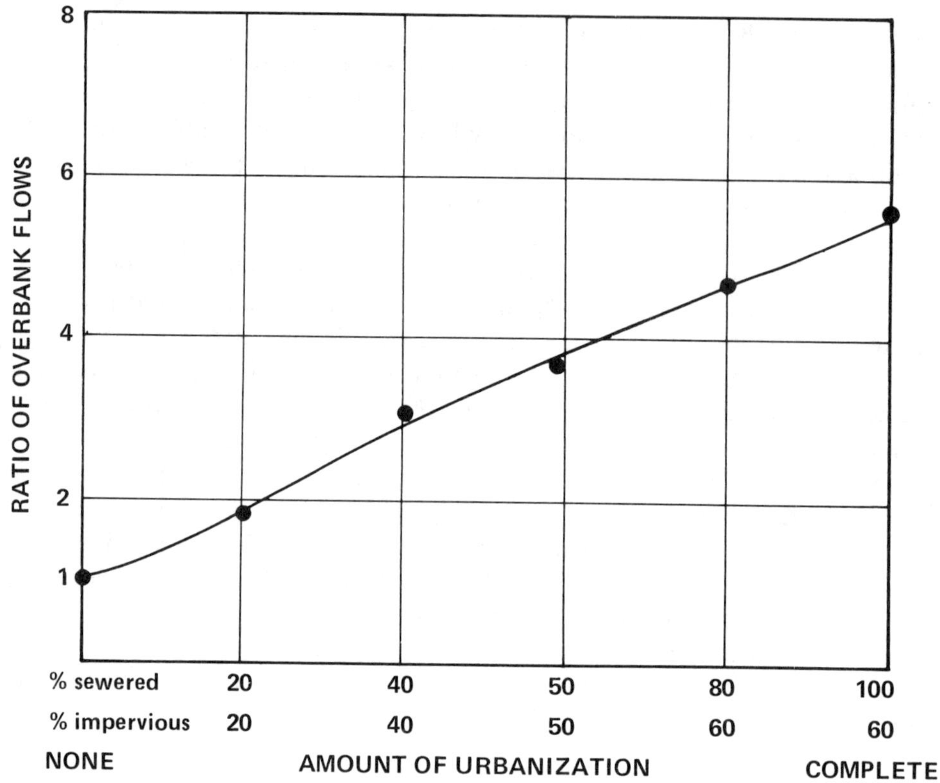

FIGURE 9-8. Increase in floods due to urbanization. Ratio of overbank flows is ratio of floods after urbanization to before. Floods are about 4 times more frequent after 50% urbanization.

FLOOD-HAZARD MAPS

A flood-hazard map shows the area of inundation resulting from either a particular flood or a probable flood (for example, a 50-year flood). The area that has been or is expected to be flooded can be approximated and plotted on a topographic map.

To construct a flood-hazard map for a probable flood at a site, determine the discharge associated with that probable flood from the flood-frequency graph. Determine the gauge height associated with the discharge from the rating curve. The gauge height may or may not be expressed in terms of elevation above sea level. If it is not, convert by adding the value of the gauge height to the datum plane. (The datum plane is usually given as feet above sea level.) The depth of flooding on the flood plain is the difference in elevation between the height of flood water (above sea level) and the elevation of the flood plain (determined from the contour lines on the topographic map). The flood boundaries are determined by extending the flooded area to the elevation on the flood plain that corresponds to the gauge height. It may be necessary to interpolate between countour lines.

As an example of the use of flood-hazard maps, let's go to the city of Chicago (Sheaffer, J. R., Davis, W. Ellis, and Spieker, Andrew M., 1970, Flood-Hazard Mapping in Metropolitan Chicago, U.S. Geol. Survey Circular 601-C). You own a piece of property along Salt Creek and have determined from the map (Fig. 9-9) that it is 23.5 miles upstream of the mouth of this

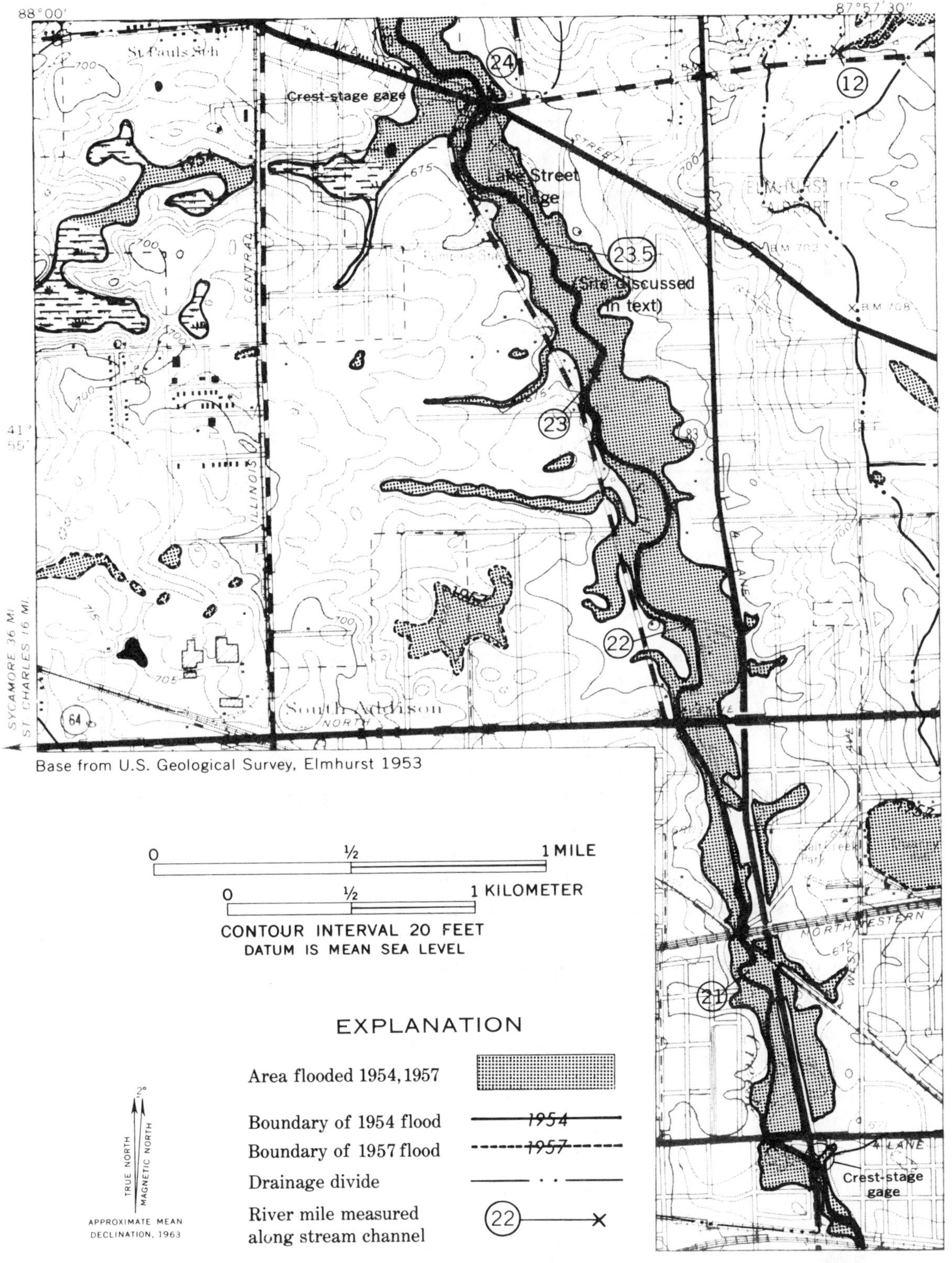

FIGURE 9-9. Flood-hazard map of the Elmhurst Quadrangle, Illinois. (From U.S. Geological Survey Circular 601-C.)

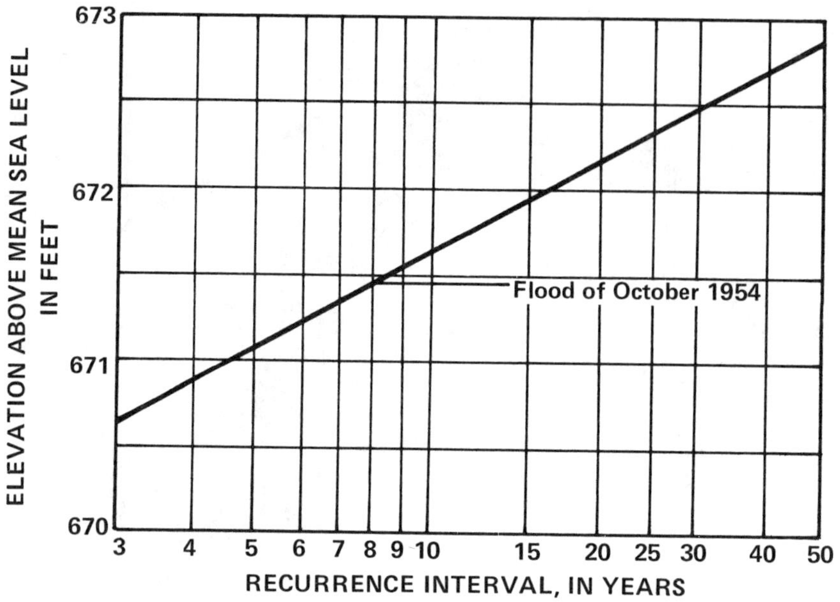

FIGURE 9-10. Flood frequency curve for Salt Creek at the Lake Street Bridge.

river. You are interested in building on the property and wish to know if it will be flooded in the foreseeable future. To answer this question go to the flood-frequency curve of Figure 9-10. Note that the 1954 flood crested at 671.5 ft at the Lake Street Bridge gaging station. However, this gage is at river mile 24 and your property is at river mile 23.5. An adjustment must be made. To do this go to Figure 9-11, which shows profiles of selected flood water elevations along the stream. Move down the 1954 profile to mile 23.5 and you'll find that the flood crested at 671 ft. Now go back to the flood frequency curve and you'll see that the 1954 flood has an 8-year return period. If you build on your property at elevation 671.0 (the 1954 flood at mile 23.5), the chances of inundation of your structure are one in eight (12½%) in any given year. These results are only probabilities; in real practice, the situation may be different. In any event the chance of inundation is greater than most of us are willing to accept. You'd better seek high ground or at least a higher position on the existing site.

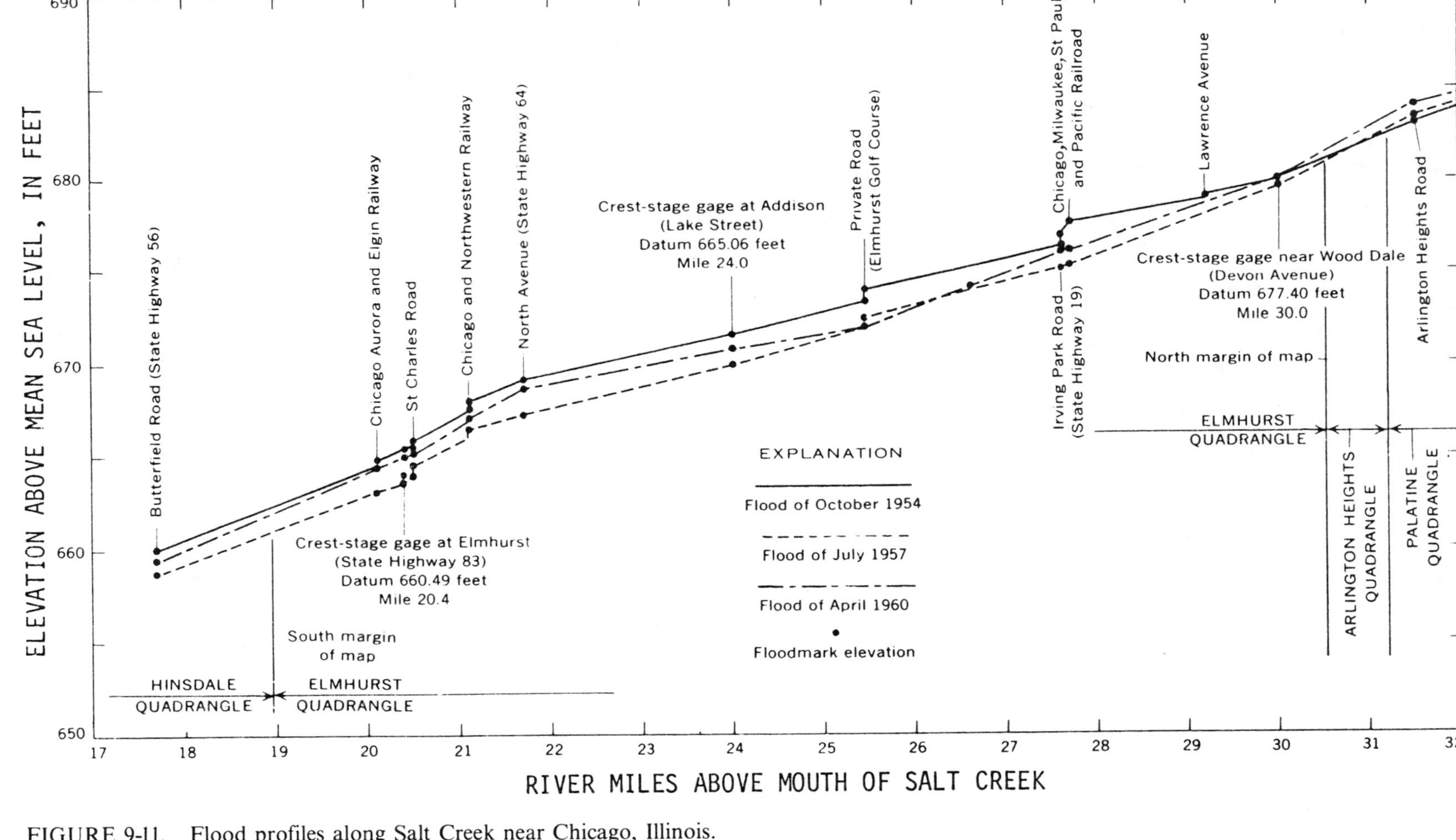

FIGURE 9-11. Flood profiles along Salt Creek near Chicago, Illinois.

EXERCISE IX

SURFACE WATER

Name ______________________________

1. Table 9-1 lists discharge amounts and gage heights for the station on the Blackstone River at Woonsocket, Rhode Island. The discharge period is from midnight August 17, 1955, through 4 a.m. August 20, 1955. The gage records flooding from a passing hurricane. Under normal circumstances the discharge-time plot of the flood would approximate a bell-shaped curve. However, you will note a sharp deviation from this theoretical curve, the result of a dam failure upstream from Woonsocket.

   (a) Plot the data from Table 9-1 on the graphs provided. Plot a discharge-time curve on the lower graph, and a stage (height)-time curve on the upper graph.

   (b) When did the flood from the dam failure reach the gage station? __________

TABLE 9-1
GAGE DATA FOR THE BLACKSTONE RIVER, WOONSOCKET, RHODE ISLAND

Date	Time	Gage height (ft)*	Discharge (cfs)
Aug. 18	2400	2.46	279
	0400	2.50	291
	0800	2.76	377
	1200	3.22	580
	1600	3.55	770
	2000	4.36	1,340
Aug. 19	2400	4.33	1,320
	0400	4.86	1,900
	0800	8.09	6,010
	1200	10.28	8,940
	1600	12.77	12,600
	2000	15.00	16,600
Aug. 20	2400	—	29,800
	0400	—	27,800

*above an arbitray datum level

   (c) Should the dam be rebuilt? Remember it was originally built to provide flood protection. Defend your answer.

____

____

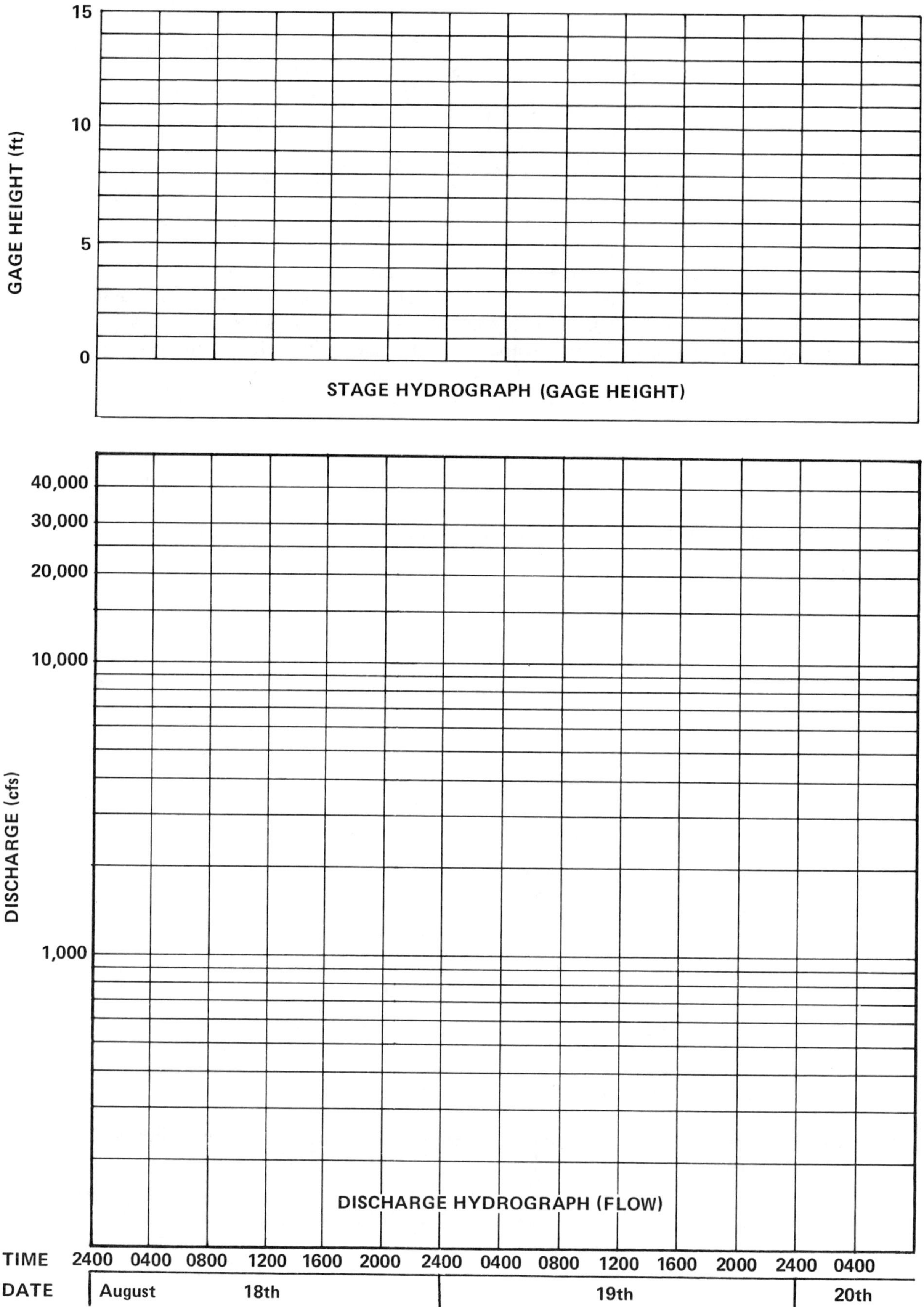

15
10
5
0
GAGE HEIGHT (ft)
STAGE HYDROGRAPH (GAGE HEIGHT)
40,000
30,000
20,000
10,000
1,000
DISCHARGE (cfs)
DISCHARGE HYDROGRAPH (FLOW)
TIME 2400 0400 0800 1200 1600 2000 2400 0400 0800 1200 1600 2000 2400 0400
DATE August 18th 19th 20th

2. Table 9-2 lists the floods (really peak discharges) for the Mississippi River at Keokuk, Iowa. The period of record is 1878 to 1952 and the table lists the discharge, relative magnitude or rank of the flow, and the return time (T) in years = (N + 1) / M.

   (a) Calculate the return time for the floods of 1888, 1919, and 1950.

   1888 ______________ 1919 ______________ 1950 ______________

   (b) On the probability paper provided, plot discharge vs. return time for the years 1878 to 1890 (13 points). Now you can see the statistical relationship between peak flow (floods if you wish) and frequency of recurrence. By extrapolating to the right, determine how often you might expect a flood of 400,000 cfs at

   Keokuk. ______________________________________

   (c) The reciprocal of return period (T) we will call probability (p). It is the chance that a flood of any given recurrence interval will occur in any year: p = 1/T. What is the probability of the occurrence of flood discharge at each of the following ranks?

   rank 1 ______________ rank 19 ______________ rank 31 ______________

3. Suppose you are building a house along beautiful Salt Creek in Chicago and are willing to accept a flood risk of once in 25 years. At what ground elevation at mile 23.5 should the building be situated (refer to Figs. 9-9, 9-10, and 9-11)? You will have to determine the 25-year flood stage (from Fig. 9-10) at the Lake Street Bridge and place a point at that elevation on Figure 9-11. Then draw a line parallel to the 1954 flood profile through the point to yield a profile along the river for the 25-year flood.

   Elevation ______________

   Do the same for the 50-year flood. Elevation ______________

TABLE 9-2
PEAK DISCHARGE FOR THE MISSISSIPPI RIVER AT KEOKUK, IOWA, 1878-1952

Year	Discharge (cfs x 1000)	Rank	T = (N + 1)/M *	Year	Discharge (cfs x 1000)	Rank	T = (N + 1)/M *
1878	150	44	1.73	1916	213	17	4.47
79	110	50	1.52	17	163	35	2.17
80	271	4	19.00	18	192	25	3.04
81	241	11	6.91	19	205	19	—
82	293	3	25.00	20	230	15	5.43
83	201	23	3.45	21	108	51	1.49
84	235	13	5.85	22	240	12	6.33
85	170	33	2.30	23	148	45	1.69
86	212	18	4.22	24	160	37	2.05
87	156	41	1.85	25	112	49	1.55
88	314	1	—	26	146	46	1.65
89	84	53	1.43	27	176	30	2.53
90	178	29	2.62	28	150	44	1.73
91	141	47	1.62	29	247	9	8.44
92	306	2	38.00	30	163	35	2.24
93	203	21	3.62	31	53	57	1.33
94	158	40	1.90	32	106	52	1.46
95	59	56	1.36	33	160	37	2.05
96	161	36	2.11	34	84	54	1.41
97	230	15	5.07	35	138	47	1.62
98	108	51	1.49	36	148	45	1.69
99	159	39	1.95	37	190	25	2.92
1900	124	48	1.58	38	193	24	3.17
01	150	44	1.73	39	159	38	2.00
02	181	28	2.71	40	82	55	1.35
03	270	5	15.20	41	154	42	1.81
04	186	27	2.81	42	201	23	3.30
05	212	18	4.22	43	174	32	2.38
06	192	25	3.04	44	255	7	10.90
07	178	29	2.62	45	203	20	3.80
08	178	29	2.62	46	223	16	4.75
09	181	28	2.71	47	246	10	7.60
10	124	48	1.58	48	234	14	5.43
11	156	41	1.85	49	151	43	1.77
12	220	16	4.75	50	176	31	—
13	169	34	2.24	51	265	6	12.67
14	122	49	1.55	52	254	8	9.95
15	142	46	1.65				

*N = total number of years in chart = 75; M = rank; T = return time.

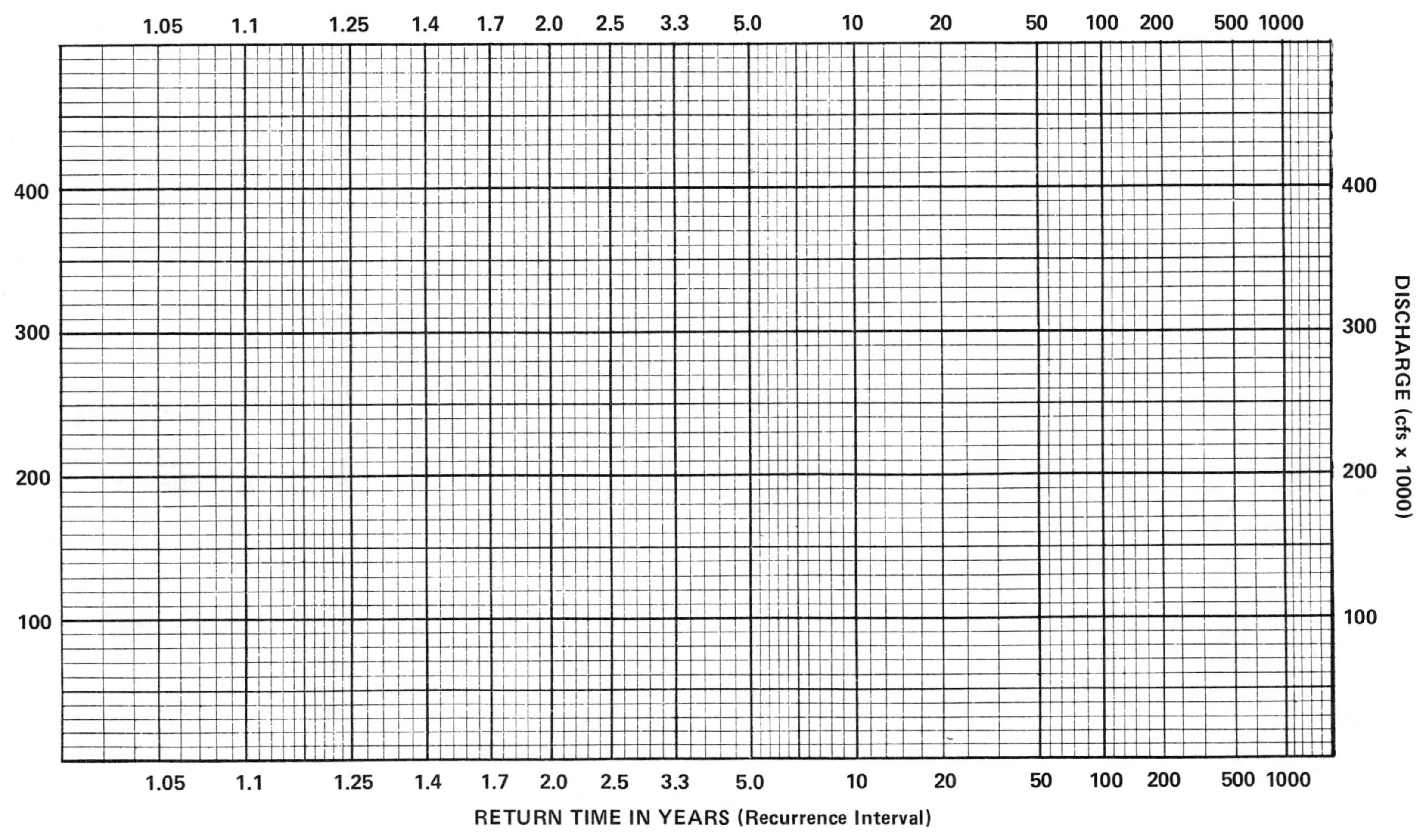
1.05 1.1 1.25 1.4 1.7 2.0 2.5 3.3 5.0 10 20 50 100 200 500 1000
400
300
200
100
DISCHARGE (cfs x 1000)
RETURN TIME IN YEARS (Recurrence Interval)

Chapter 10
Subsurface Water

As we have already learned, some of the water that falls on the Earth seeps directly into the ground. But water also flows along the surface of the Earth in streams or collects in reservoirs, lakes, and oceans, and some of this surface water eventually infiltrates the ground. Hence, there is a direct relationship between the amount of surface water and the amount of water in the soil and rock just below the Earth's surface (Fig. 10-1). And like surface water, ground water can be beneficial, as when it supplies our needs for consumption, irrigation, or industry, or non-beneficial, as when it promotes landslides, seeps into tunnels, damages building foundations, or otherwise poses a hazard.

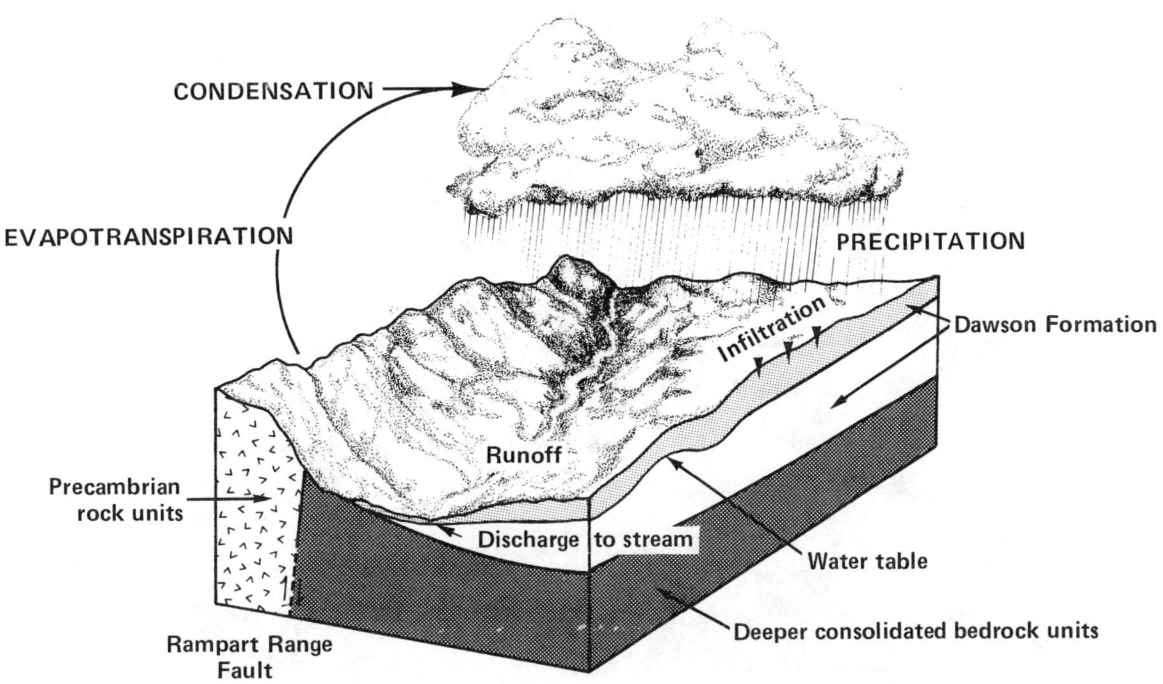

FIGURE 10-1. The hydrologic cycle in El Paso County, Colorado. (From U.S. Geological Survey Professional Paper 950, 1978.)

In seeking to minimize the hazards of subsurface water, the environmental geologist is especially interested in studying such variables as depth to water table; direction, velocity, and amount of discharge of flow; ground water withdrawal and storage; and sources of pollution. In the sections that follow, we will discuss some of these.

INFILTRATION AND PERCOLATION

As water infiltrates the ground, it moves through a thin, moist zone of weathered soil and rock. The water continues its downward migration through a region known as the **zone of aeration** where the pore spaces contain either water or air. Ultimately, the water reaches a **zone of saturation** where the pore spaces are completely filled with water. The water in the zone of saturation is called **ground water**, and the contact between the zone of aeration and the zone of saturation is called the **water table** (Fig. 10-2). Water may be trapped above the main water table in certain locales. These **perched ground water zones** exist because infiltration is inhibited by local impervious rocks or soil layers. In fine-grained rocks, water may rise above the water table by capillary action; this zone is called the **capillary fringe**.

The water table commonly is an irregular surface that reflects in a general way the overlying topography or bedrock configuration. The high areas of the water table are called **ground water divides**. Flow is away from such divides.

AQUIFERS AND AQUICLUDES

The zone of saturation may be found in the sand and gravel (alluvium) of a stream bed, flood plain, or deep basin. It may also exist in any permeable rocks, such as sandstones,

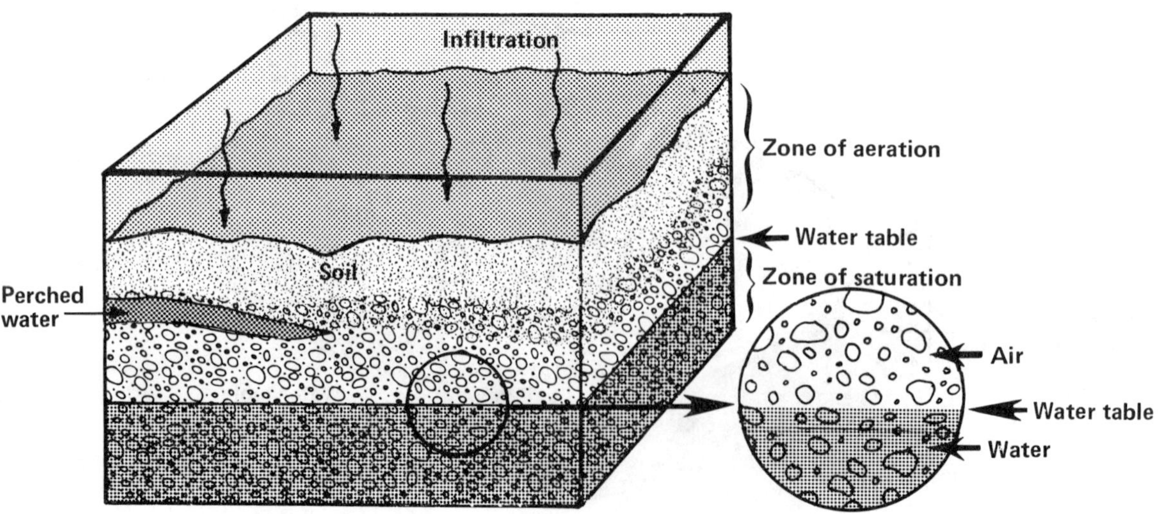

FIGURE 10-2. Ground water zones.

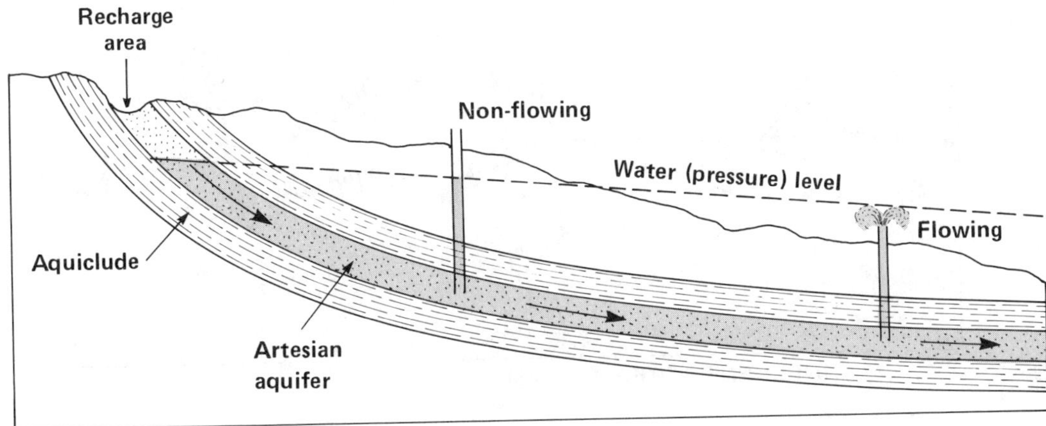

FIGURE 10-3. Confined aquifer showing flowing and nonflowing artesian wells.

limestones, and basalts. Ground water moves through connected pores between individual grains or through fractures and joints in a rock. The lateral flow of water is called **percolation**. If the soil or rock contains sufficient ground water that can be released in useful quantities, the soil or rock is called an **aquifer**. If it is impermeable, it is called an **aquiclude**. The distinction is a function of the permeability of the earth material.

Aquifers may be **unconfined** or **confined**. An unconfined aquifer is one that is free from overlying and continuous impermeable layers, whereas a confined aquifer has an impermeable layer overlying it (Fig. 10-3). The confined aquifer may contain water under pressure from this overlying layer. If the confining layer is pierced by a drill, the water in the confined aquifer may rise above the confining layer and possibly even reach the ground surface. Such confined aquifers are called **artesian systems** and the wells are called **artesian wells**. The height to which water rises in an artesian well depends upon the amount of pressure in the system.

RELATION OF STREAMS TO WATER TABLE

Streams can be divided into two types, depending on whether they are located above the water table or at the water table. A stream located above the water table is called an **influent stream**. This type of stream gets its water entirely from surface runoff. Water from influent streams feeds the ground water (Fig. 10-4). Locally, the infiltrating water may build a mound of water above the general level of the water table; such a mound is called a **recharge mound**. A stream located at the intersection of the water table and the ground surface is called an **effluent stream** (Fig. 10-4). This type of stream gets its water from both surface runoff and ground water.

There is a practical reason for being able to recognize and distinguish between these two types of streams. The waters of the effluent stream may not be suitable for human consumption where the local ground water is polluted.

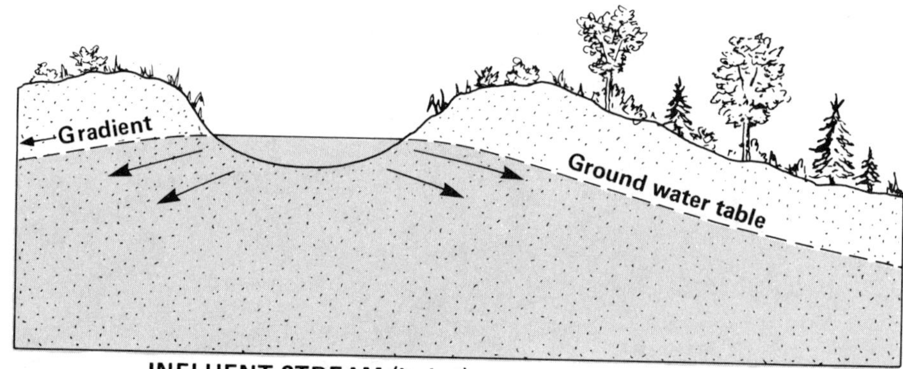

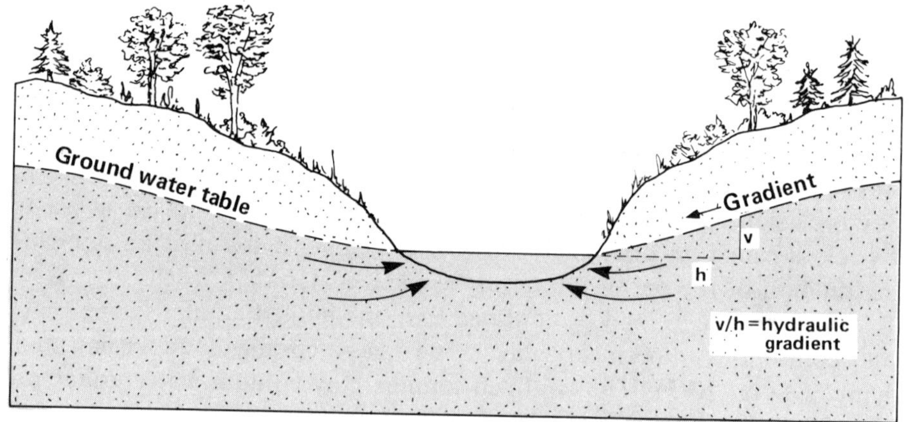

FIGURE 10-4. Streams that are gaining or losing water.

MOVEMENT OF GROUND WATER

Ground water, like surface water, flows toward the ocean. Whereas surface water flows relatively rapidly in well-defined channels (stream beds), ground water has no such channels and moves slowly.

The direction of ground water flow can be determined by the slope of the water table. (Remember, the water table is the top of the ground water zone.) The slope and shape of the water table can be drawn from data about water elevations measured in wells or drill holes. These elevations can be used to make a contour map of the water table in the same way that surface elevations are used to make a topographic map (Fig. 10-6). The direction of ground water flow is toward the lower elevations, approximately perpendicular to the contours representing the water table. The slope of the water table provides a means for measuring the **hydraulic gradient** of the ground water surface. The method is similar to that used in determining the slope of a river (see Topographic Maps, Chapter 2). If the water table slopes one foot vertically in 100 ft horizontally, the hydraulic gradient is 0.01.

The discharge velocity, V, of ground water flow is proportional to the hydraulic gradient, S, and to a coefficient of permeability, k, (measured in ft/day or m/day).* This relation is known as **Darcy's Law** and is

$$V = Sk$$

The velocity of flow is expressed in the same units as the coefficient of permeability. For example, if k is 50 ft/day and S is 0.01, the velocity is 0.5 ft/day. The discharge, Q, is the volume of water passing through an area, A, of one square foot and is

$$Q = VA$$

In the example cited above, if the velocity is 0.5 ft/day and flows through one square foot, the discharge is 0.5 ft^3/day.

GROUND WATER WITHDRAWAL

In order for the ground water to be useful it must be pumped to the surface. The effects of withdrawing water from the subsurface may not be entirely beneficial. The water table is lowered, the pattern of ground water flow may change, and **subsidence** of the land surface may result. In coastal areas, withdrawal of the ground water may result in **sea-water intrusion.** To a great extent, the disadvantages may be mitigated or avoided by proper management, such as controlling the amount of withdrawal or artificially recharging the ground water system with water transported to the area.

As water is pumped from the underground reservoir, the water in the immediate vicinity of the well flows toward the well. This results in a local gradient toward the well and the water table is locally lowered. The shape of this lowered water table is called a **cone of depression** (Fig. 10-5) and the distance that the water level decreases is called the **drawdown.** If many wells are spaced closely and the water is pumped, the resulting cones of depression would merge and change the shape of the water table, with attendant changes in the hydraulic gradient.

Since subsurface water exercises a buoyant influence on rocks and soils, as water is pumped from the aquifer, the fluid pressure decreases, buoyancy decreases, and the apparent weight of the rocks and soil increases. The grains of rock or soil respond to the change of pressure and **compaction** results. This compaction has three effects: (1) it lowers the ground surface, (2) it decreases the permeability and water discharge, and (3) it may result in **subsidence** of the ground surface.

---

*There are two uses for the term "coefficient of permeability." The one described here is used by engineering geologists and is expressed in terms of velocity. Hydrologists express the coefficient of permeability in terms of discharge and use the symbol Kp; it is the rate of flow, in gallons per day, through a cross-sectional area of 1 square foot under a hydraulic gradient of 1 foot per foot at a temperature of 60° F. The unit of flow is called the Meizner unit (in the United States); elsewhere in the world it is expressed as "cubic meters per day square meters" (m^3/dm^2).

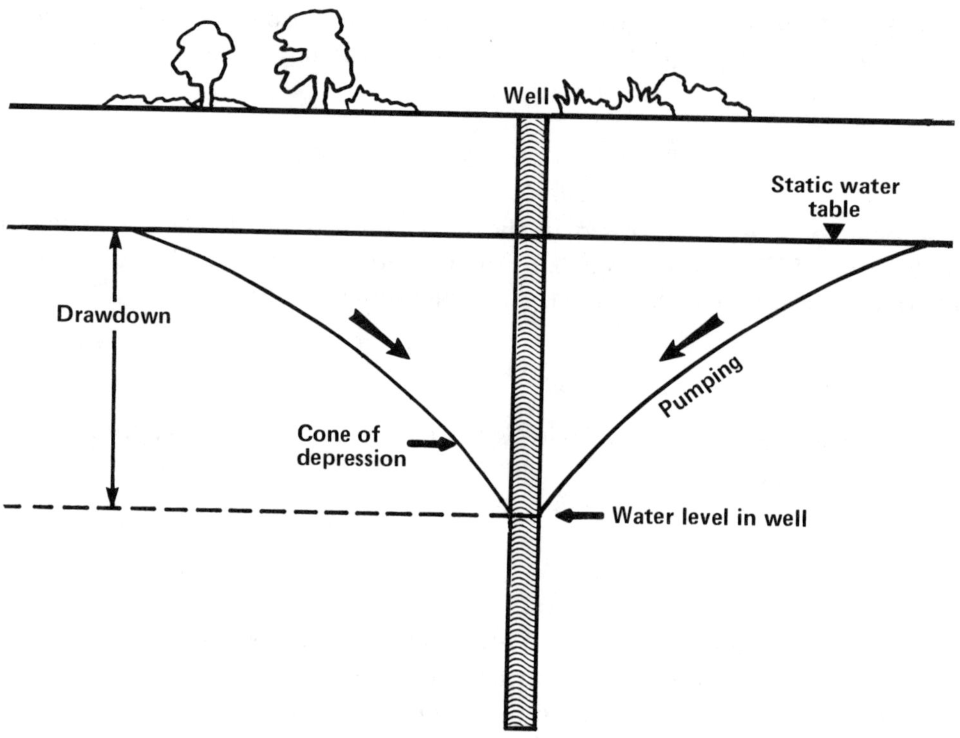

FIGURE 10-5. Cone of depression and drawdown in a pumping well.

In coastal areas, the withdrawal of ground water may result in intrusion of salt water. Ordinarily, the fresh water in the aquifer inhibits the salt water from intruding. This is known as the **Ghyben-Herzberg effect**. Sea water is 2.5% heavier than fresh water because of its salt content and fresh water will float on sea water, although some mixing of the waters occurs at the boundary. The fresh water depresses the salt water, and over a long period of time, will drive the salt water out of the aquifer. As long as the hydrostatic equilibrium is maintained by the normal flow of fresh water toward the ocean, the fresh water acts as a barrier to salt water encroachment. However, if pumping of the fresh water occurs, the hydrostatic equilibrium is disturbed and as the fresh water is diminished, the salt water will move below the fresh water. The salt water may eventually reach the water wells and be pumped to the surface. In areas where the fresh water is used for domestic purposes or for irrigation, such use may have to be curtailed when the salt water being pumped is deleterious.

EXERCISE X

SUBSURFACE WATER

Name __________________________________

The area of Gillette, Wyoming, is being considered for open-pit mining for coal. Figure 10-6 shows the ground water level in the general area. The levels of ground water can be used to determine the direction of ground water flow and the relative spacing between contour lines gives some idea as to the shape of the water table (using the same principles as for topographic maps).

1. At the points marked X, draw arrows one inch in length to show the direction of ground water flow (——————►). The map has 21 points.
2. Draw a dashed line along the ground water divide (along the crest line of the contour map), thereby defining the area of generally consistent orientation of subsurface flow toward a water basin. The arrows drawn in Problem (1) should be consistent within the flow area.
3. If a waste-water injection well were planned at A, would the water flow toward Gillette? Explain.

   ____

   ____

4. Because of economic considerations, the City Council decides to put in a waste-water injection well one mile outside the city limits. You are asked to locate the injection well. Where would you put it and why?

   ____

   ____

   ____

5. What is the value of the steepest hydraulic gradient in the Gillette area? __________
6. What is the general direction of ground water flow? ____________________

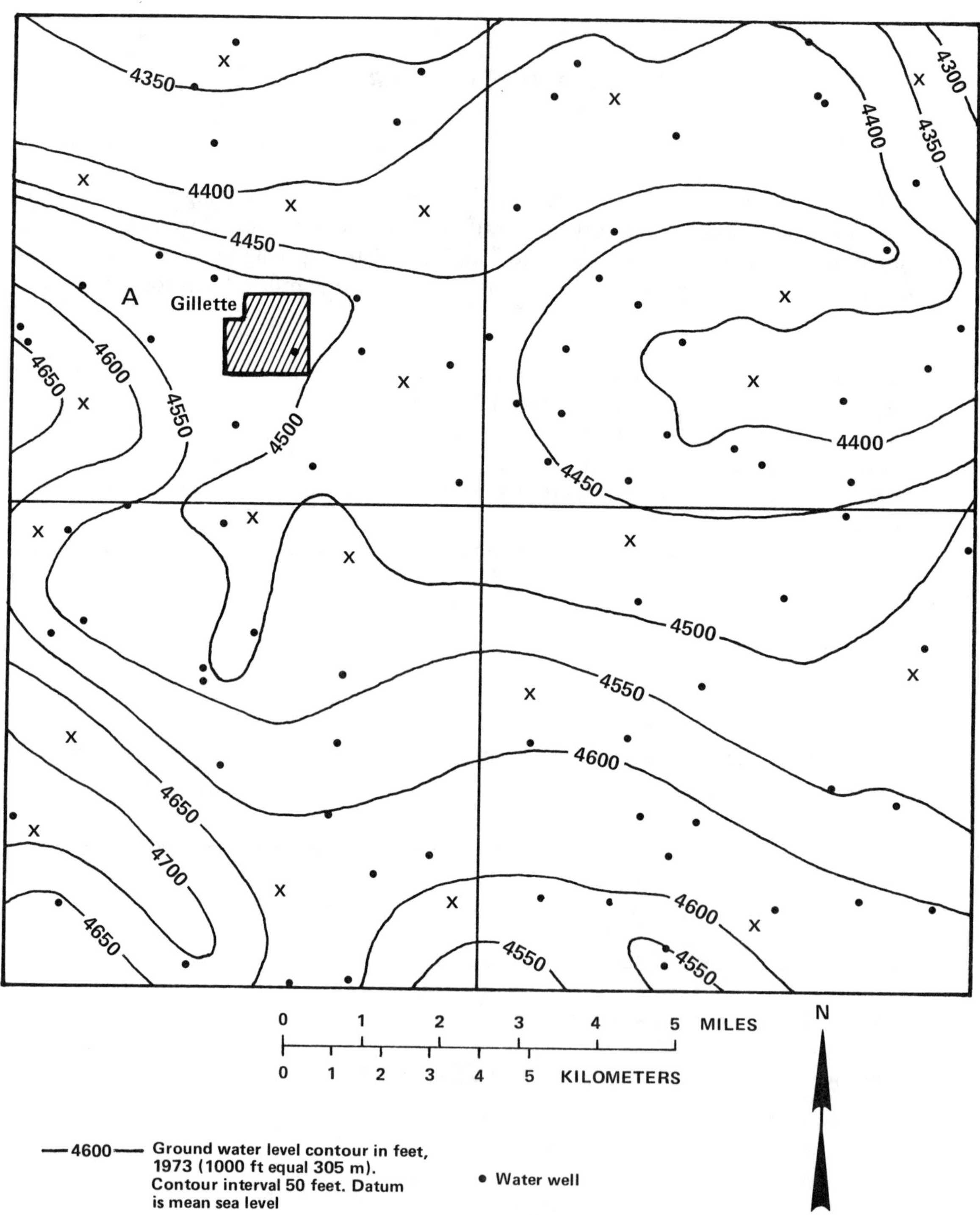

FIGURE 10-6. Ground water levels in the vicinity of Gillette, Wyoming.

Chapter 11
Coastal Processes

Beaches are dynamic. They change rapidly in response to waves, tides, or interference by man. The average citizen views the beach as a valuable recreational resource, but to the geologist it represents the last rampart protecting land from attack by marine processes. When this resource is lost, regardless of one's view, there follows property damage and loss of esthetic amenities. Therefore, any steps to stabilize our beaches must be very carefully considered in advance.

CLASSIFICATION OF BEACHES

Our classification of beaches is based on land forms surrounding the coastal zone. These land forms affect beach dynamics through their influence on the near-shore processes that dominate at the site.

Coastal Plain Beaches. These beaches are backed by vast coastal plains and lowlands such as those along much of southern California. They are subject to wave attack from many directions and from open ocean storms (Fig. 11-1).

Barrier Beaches. Also known as offshore bars, barrier beaches are found along coastlines with flat, near-shore shelves and a large supply of sand, as along the Atlantic and Gulf coasts of the United States. They are subject to direct attack by large waves and, more importantly, to high surge and inundation by hurricanes. Atlantic Beach, New Jersey, and Miami Beach, Florida, are built on offshore bars (Fig. 11-2).

Spits. Spits are long fingers of sand attached to the mainland, usually across the mouth of a bay or inlet. They are built by currents where there is a large sand supply (Fig. 11-3). Cape Cod, Mass., is an example.

Pocket Beaches. These beaches are found along cliffed shorelines where coves are found. They are generally well protected and represent somewhat closed systems (Fig. 11-4).

BEACH PROCESSES

Beaches may be composed of almost any granular material, such as gravel, sand, or mud; however, those beaches of greatest recreational value are composed of sand-size

FIGURE 11-1. Coastal plain beach, Huntington Beach, California (U.S. Navy photograph).

material (1/16 mm to 2 mm in grain size). Sand is generated in nature by the weathering of crystalline rocks, such as granites and gneisses, and the release of quartz and feldspar grains in upland regions that drain to the sea. Thus we see that beaches derive not from offshore materials, nor from the marine weathering and erosion of local rocks, but from rock materials further inland.

Beach sand is carried from the land to the sea by rivers. The mouths of these rivers act as point sources for beach material and waves are the prime movers or conveyor belts that distribute this material along the beach. Waves are generated by winds at sea both locally and from distant sources. These waves travel from their origin to the coastline and dissipate their energy as surf or breakers. In the surf zone, a wave changes from one where the water simply oscillates in a circle to a **wave of translation**, whereby the water actually moves forward and is discharged against the shoreline. Sediment is thus put into suspension and transported along the coastline. Because waves do not usually approach parallel to the shoreline, a weak current is generated along the beach known as a **longshore** or **littoral current** (Fig. 11-5). The strength of this current is a function of the angle of wave approach and is proportional to the sine of the angle that the wave makes with the shoreline (sine 0° = 0, sine 90° = 1). This longshore current is responsible for carrying bathers and sand down-current. The beach material transported by the current is known as **littoral drift**. Rates of littoral drift vary greatly with wave energy and angle of

FIGURE 11-2. St. Vincent Island, a barrier beach off the coast of Florida. The grooved appearance is actually due to storm-built beach ridges and intervening low marsh lands. (U.S. Geol. Survey photo.)

FIGURE 11-3. A spit has developed over the mouth of an abandoned estuary, Newport Beach, California (U.S. Navy photograph).

approach, but range from a few thousand cubic yards to more than 1,000,000 cubic yards per year. To give an idea of the magnitude of this transport, a cubic yard of sand is roughly equal to one square foot of beach. If you wish to describe this in monetary terms, the cost to place a cubic yard of sand on a beach would be about $3.00 in 1980 dollars.

In addition to beach drifting, sand also moves on and offshore with seasonal variation in waves. During the winter most shorelines experience high waves, and sand is eroded from the beach and stored in sand bars (Fig. 11-6). The beach is narrowest at this time. With the advent of summer surf conditions and low waves, the sand in storage is gradually worked shoreward and the beach is built out. These annual cycles of cut and fill are common to all beaches.

COASTAL HAZARDS

Natural hazards of greatest importance along the coast are: (1) tropical cyclones or hurricanes, (2) tsunami, and (3) beach erosion. Because the coastal zones tend to be heavily populated, the impact of these hazards is great. In the United States, 12 of the 13 largest cities are located in the coastal zones and approximately 75% of our population is in coastal states.

Tropical Cyclones. These include hurricanes (Atlantic terminology), typhoons (Pacific terminology), and other storms of lesser wind velocities that claim many lives and cause enormous amounts of property damage yearly. During hurricanes, water is known to rise along the shoreline to elevations far in excess of normal high tides and much of the land area is inundated. This is particularly true of the west coast of Florida and the Gulf Coast states. For example, Hurricane Allen struck the Texas coast at about 8:00 a.m. on August 9, 1980, with winds up to 260km/hr (150 mph). On the satellite photo showing its northwesterly track across the Gulf of Mexico it is a great pinwheel. Winds of 300 km/h were recorded and hurricane force winds extended 240 km from its center (Fig. 11-7). In its track across the Gulf it crossed South Padre Island, a barrier beach, and did considerable damage to this resort community. The island is 177 km long but only a few kilometers wide, and has very low relief. For the most part the island is only 2.5 m above sea level, except for wind-blown dunes up to 12 m high. Between the barrier island and the mainland is Laguna Madre, about 10 km wide and averaging 3 m deep. A storm surge about 3 m (above stillwater tide level) was generated by the sustained winds of 300 km/hour.

A study of damage by geologists at Texas A & M University showed that streets on the island oriented perpendicular to the beach created low passes for water to flow through the dune area. Beach houses, condominiums, and resort hotels on these islands were built with deep foundations or on pilings so that wave action and erosion, or scour, would not undermine the structures. Retaining walls to deflect storm surge were also an integral part of

FIGURE 11-4. Pocket beaches in cliffed shoreline near Laguna Beach, California (U.S. Navy photograph).

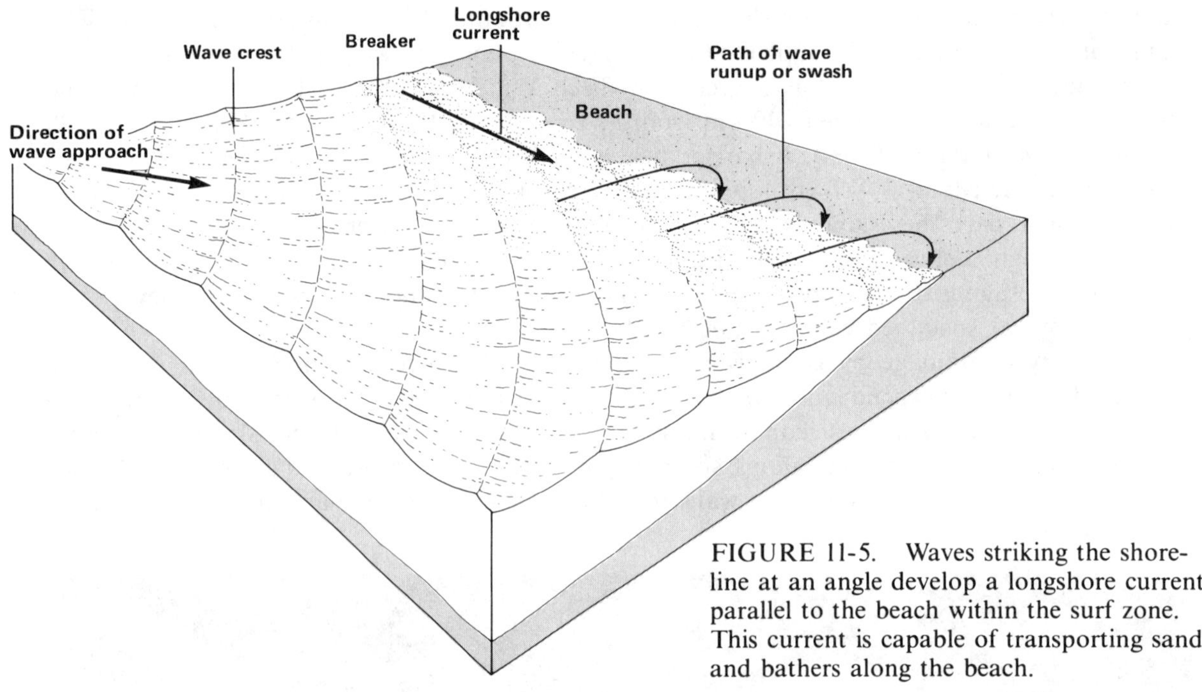

FIGURE 11-5. Waves striking the shoreline at an angle develop a longshore current parallel to the beach within the surf zone. This current is capable of transporting sand and bathers along the beach.

the newer buildings. Most of the surge damage occurred to beachfront retaining walls. They failed because scouring at the base removed sand fill from behind the walls and they overturned (Figs. 11-8 and 11-9). Storm waters crossed through the island in low areas such as old washover channels (from previous hurricanes) and paved streets; the water damaged older houses.

A major conclusion of the study of hurricane damage is that old washover channels and roads across barrier islands will be routes of concentrated flow and wave erosion. Also, large, well-built structures with seawalls will channelize and increase flow to the back of the island. Planning should take into account this increased flow and the fact that structures without retaining walls or seawalls should be considered "disposable" (Gowan, Samuel, and Gregg, Jack, 1980, Hurricane Allen Special Update; Assoc. Engineering Geologists Newsletter, pp. 22–26).

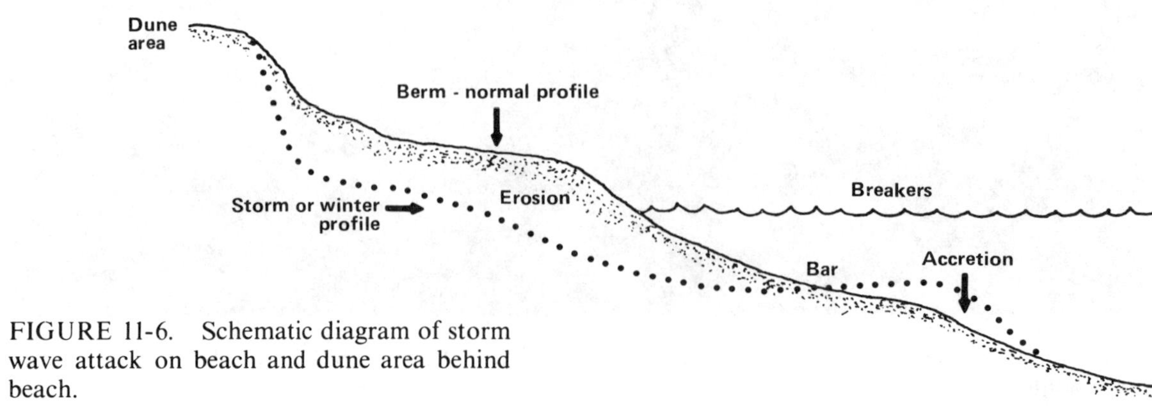

FIGURE 11-6. Schematic diagram of storm wave attack on beach and dune area behind beach.

FIGURE 11-7. Hurricane Allen in the Gulf of Mexico, August, 1980. (Courtesy Sam Gowan and Jack Gregg.)

Tsunami. Seismic sea waves, or tsunami, are particularly hazardous around the Pacific Ocean basin. They are caused by sudden displacements of the sea floor along faults, usually in trenches, and move with high velocities across the open oceans to wreak havoc on unprotected shorelines. They are most common in the Pacific, less so in the Indian Ocean, and rare in the Atlantic. Tsunami may also be caused by submarine volcanic eruptions, such as the one-hundred-foot wave resulting from the Krakatoa eruption of 1883 in the Indian Ocean (see Chapter 7), and submarine landslides, such as the one that destroyed the little city of Valdez in the great Alaskan earthquake of 1964. In open waters these waves travel with velocities as great as 600 mph with a wave length (the distance between crests) in excess of 100 mi. Because wave heights in deep water are only a foot or two, waves go unnoticed. When waves enter shallow coastal waters, wave velocity decreases, wave length decreases, and the wave height increases greatly. Tsunami as high as 100 ft have been reported; however, they are usually only 20–30 ft high. Damage is the result of runup inland and is particularly extensive in low-lying coastal areas. Damage is minimal along steep or cliffed coastlines because wave energy is reflected. Although at present we cannot predict tsunami, we can detect them and warn inhabitants of coastal communities exposed to their destructive forces (Fig. 11-10).

FIGURE 11-8. Damage to condominiums and beach cottages from Hurricane Allen. Note damaged retaining wall. (Courtesy Sam Gowan and Jack Gregg.)

FIGURE 11-9. Storm-surge damage from Hurricane Allen. (Courtesy Sam Gowan and Jack Gregg.)

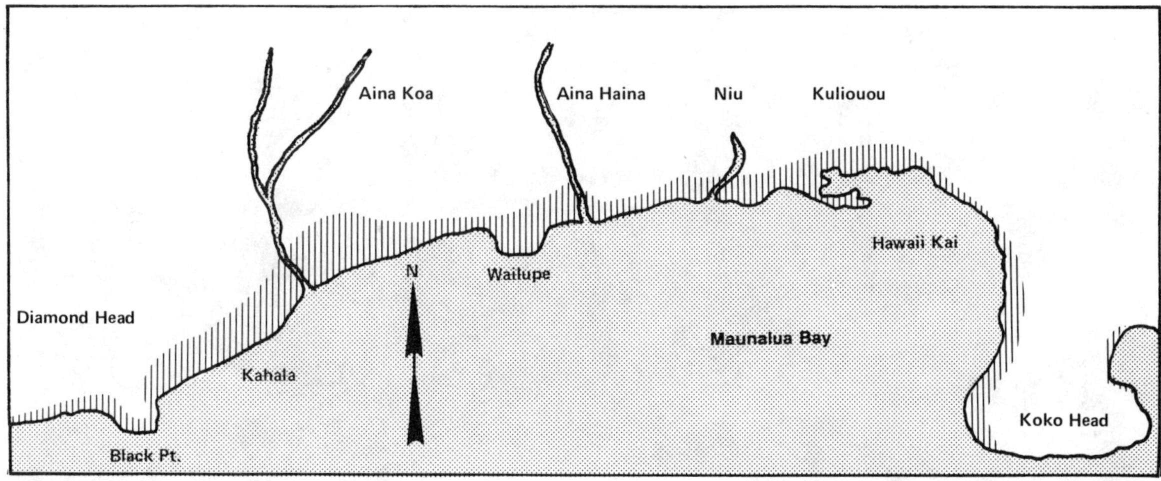

FIGURE 11-10. Tsunami runup on the island of Oahu near Honolulu. (From Honolulu Telephone Book.)

Tsunami damage may be minimized somewhat by knowing the direction of wave runup and designing structures accordingly. Studies from the 1960 tsunami at Hilo, Hawaii, revealed the direction of runup from the orientation of bent parking meters. It was then recommended that future structures in the area be built with their long axis parallel to the direction the wave is traveling to reduce the effect of the wave's force. Studies also indicated that structures with raised or open first floors fared much better than structures with walls solid to the ground.

Coastal Erosion. In contrast to erosion produced by a tsunami and hurricane, normal beach erosion is generally a slower and more continuous process. It is measurable and predictable and when not controlled has caused millions of dollars in property damage. Erosion occurs when the supply of sand to a given area is less than that removed by wave action and longshore currents. In contrast, when supply exceeds removal, beaches widen. The sand is blown inland by the wind and coastal dunes are created, usually near rivers that supply enormous amounts of sand.

The distribution of river-supplied sand by longshore currents is a natural process. Where supply and removal are in balance, stable beaches exist that are neither growing nor eroding. However, man's interference with natural shoreline processes has been generally detrimental. The supply of sand to beaches has been reduced by dams and flood-control projects in coastal watersheds. In addition, urbanization has resulted in paving over much of the watershed that was the source of granular material. The existing sand is continually being removed by wave action but the replacement supply from local streams and rivers has been drastically reduced. The beach gradually retreats. To prevent property loss, a variety of seawalls or rock revetments is built to dissipate or reflect wave energy. The beach is no longer usable and a valuable resource has been lost.

Local erosion and accretion of sand can result from damming littoral drift with a newly constructed groin or jetty (Fig. 11-11). Sand accumulates updrift of the groin, and erosion occurs downcurrent. The extent of accretion and erosion depends on the season and the length and type of the groin. As soon as the wedge of sand reaches the end of the groin,

FIGURE 11-11. Jetty and groin at Seal Beach, California. Note sand accretion on northwest side of structures and erosion on opposite side. Extensive erosion occurred on the downcurrent side of the long jetty requiring dredging of sand from the harbor to fill downcurrent beaches and protect homes.

drift will occur around the structure and erosion will no longer occur downcurrent. Also, some groins are constructed to be permeable so that some sand can work its way through the structure and decrease the amount of erosion downstream. Littoral currents can move in both directions along the beach and it is only where there is a dominant direction that accretion and erosion become a problem. For instance, in southern California the predominant drift direction is southerly, even though there are many days of the year when swell comes from the south and the longshore current moves in the opposite direction.

In many cases where extensive erosion has resulted from construction of a groin or breakwater, such as at Santa Barbara, California, man has installed dredges and pumps to remove sand from the accretion area to the areas of erosion. This process is called **bypassing** and is now designed into most large coastal engineering works where littoral drift will be impeded. Finally, in areas of critical erosion, sand has been imported to artificially nourish the beaches. This is expensive but necessary in many regions. The present problem is that sand as a resource is becoming more valuable and suppliers are reticent to sell large volumes (millions of cubic yards in some cases) for beach protection when it is so badly needed by the building industry.

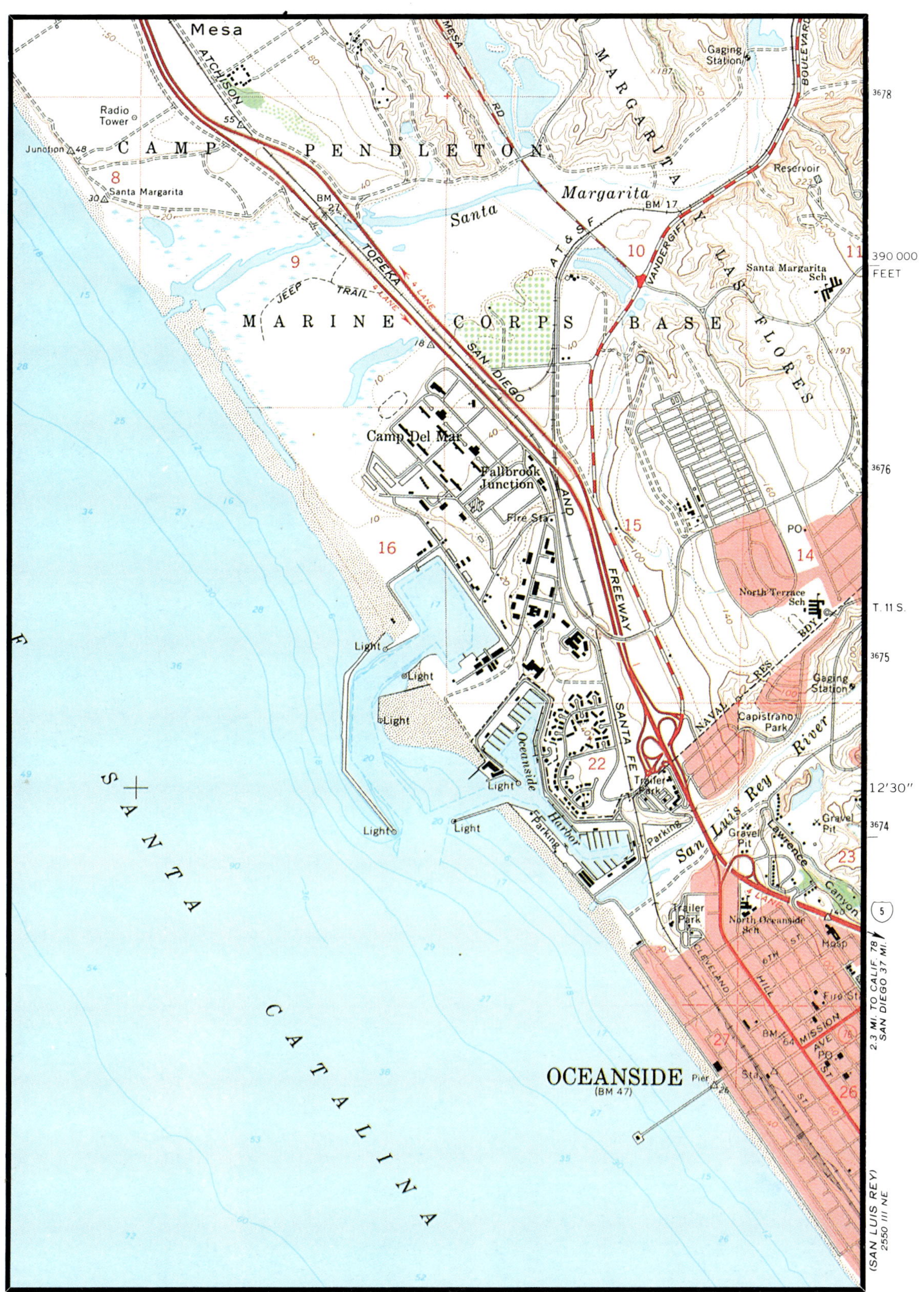

FIGURE 11-12. Portion of the Oceanside Quadrangle, California. Scale 1:24,000.

EXERCISE XI

COASTAL PROCESSES

Name ______________________________

1. In recent years, an increasing number of streams in the Los Angeles area have been blocked by dams. What effects, if any, will these structures have on the beaches down-current from the mouths of these streams?

____

____

____

2. Study the series of air photos of Sandy Beach near Atlantic City, New Jersey (Fig. 11-13).

   (a) What coastal feature has developed here and why? ______________________________

   ____

   (b) What is the apparent direction of longshore drift? How do you know? __________

   ____

   (c) What is the probable direction of wave approach? ______________________________

   ____

3. Examine the map of Oceanside, California (Fig. 11-12, p. 163). On the map, indicate the following:
   (a) Direction of wave approach.
   (b) Direction of longshore drift.
   (c) If a series of three groins were built along the shoreline, what would be the patterns of sand accretion and erosion? (Draw in the groins and predicted depositional patterns.)
   (d) Why hasn't a feature similar to that illustrated by Sandy Beach (Fig. 11-13) developed at Oceanside, California?

   ____

   ____

   ____

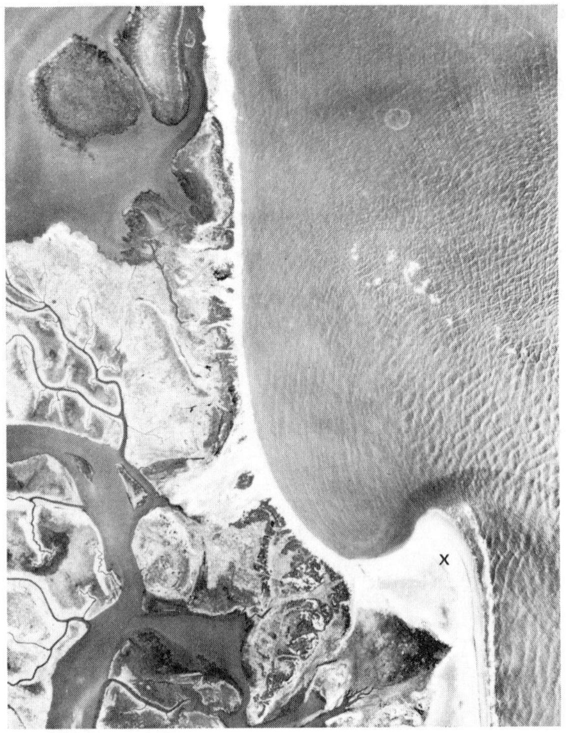

a.

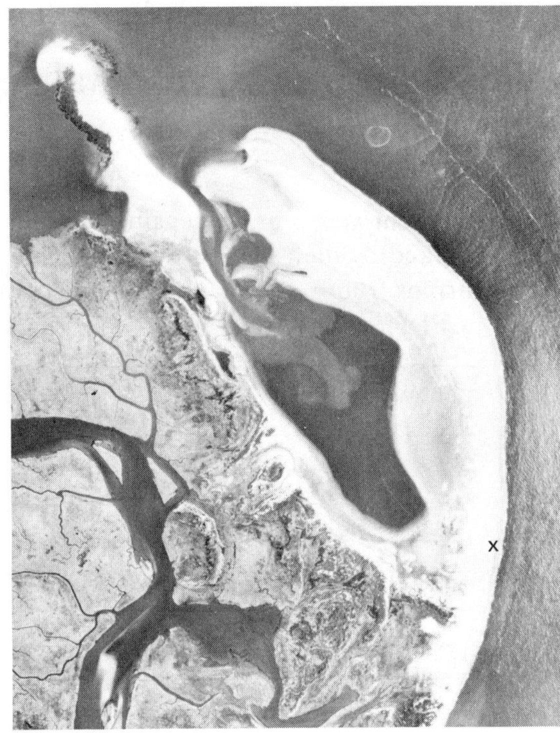

b.

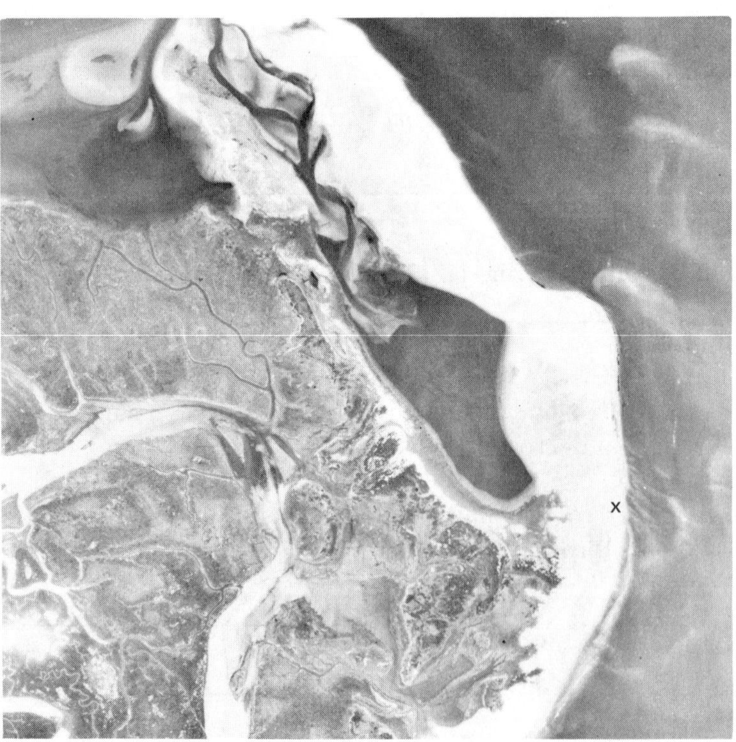

c.

FIGURE 11-13. The natural transport of sand over a 23-year period at Sandy Beach, New Jersey. (a) First stage of development in 1940. (b) Between 1940 and 1957 the spit extends from point "x". (c) 1963. The spit joins the land and another spit starts to develop.

4. Assume you have recently acquired property on a relatively narrow beach in Florida. Your neighbor (updrift) wishes to build a groin to widen the beach in front of her house.
   (a) Is the groin in your best interest? Explain.

   ____

   ____

   ____

   ____

   ____

   (b) If your neighbor does build a groin, what might be your reaction to preserve the value of your property? ______________________________

   ____

   ____

   (c) Sketch this hypothetical case.

5. Bolinas, a small town in California, has a problem with coastal erosion. The rate of seacliff retreat is approximately 1.5 ft per year under natural conditions; the rate is somewhat less when seawalls and other protective measures are used to protect the cliffs.

   Four alternatives have been considered to reduce the impact of coastal erosion in Bolinas. These have been listed in Table 11-1. A map of the Bolinas area is provided (Fig. 11-14). Using all the information, answer the following questions.
   (a) On the map, show where the shoreline will be in 50 years based on current erosional rates. Also show where the shoreline will be in 100 years.
   (b) Give the advantages and disadvantages of the proposed alternatives in the boxes provided in Table 11-1.

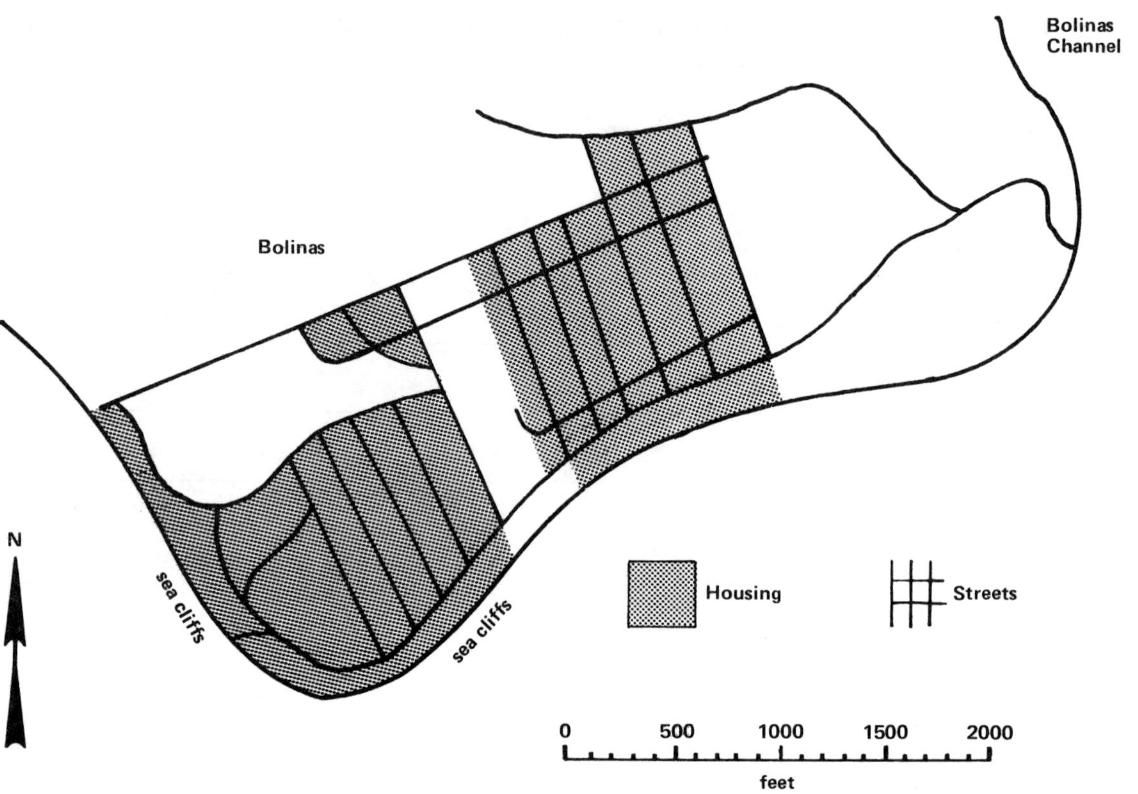

FIGURE 11-14. Idealized drawing of shoreline conditions at Bolinas, California. Not drawn to scale.

(c) What alternative or combination of alternatives would you recommend and why?

____

____

____

____

6. (a) What is a tsunami? ______________________________

____

(b) Why do tsunami occur predominantly in the Pacific Ocean and not in the Atlantic Ocean? ______________________________

____

____

TABLE 11-1
WAYS TO REDUCE COASTAL EROSION AT BOLINAS, CALIFORNIA

Alternatives	Advantages	Disadvantages
1. Zoning to prevent further development in hazardous areas. (No structures are proposed to prevent erosion.)		
2. Public land acquisition in hazardous areas. (No structures proposed to prevent erosion.)		
3. Construction of a seawall at an estimated cost of \$3–4 million. (Will not prevent continued erosion at the top of slope, but will prevent excessive erosion at base of slope.)		
4. Combination of groins, beach fill, and energy dissipators at an estimated cost of \$4–6 million. (Will not prevent some erosion at top of slope, but will prevent further erosion at base of slope. Will also serve to widen the present beach area.)		

(c) What can be done to lessen the damages incurred in tsunami-prone areas?

____

____

(d) Using the travel-time diagram of Figure 11-15, determine the time for a tsunami originating in the Aleutian trench area to arrive at the following locations.

Hawaii ____

Valparaiso, Chile ____

What is the average velocity of the tsunami in its travel time to Hawaii?

____

What is the average velocity of the tsunami in its travel to Valparaiso?

____

Give a rational explanation for the difference in velocity between the two stations.

____

____

____

____

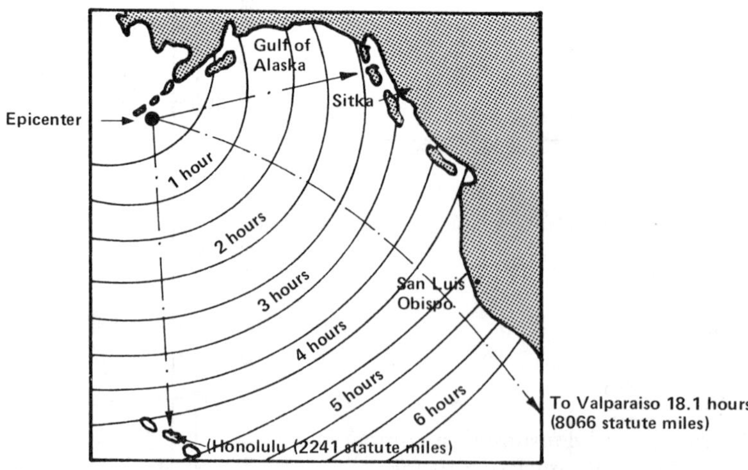

FIGURE 11-15. Tsunami travel time from a distant source in the Aleutian trench.

Chapter 12

Water Resources

Water pervades our lives. It is the most remarkable of common liquids and chemical compounds. Our bodies are 70% water and we must have a quart and a half each day to survive. We play on water, in it, and over it. Any land development, no matter what its purpose, must consider the acquisition of water, its use, and its disposal.

DISTRIBUTION OF WATER ON EARTH

Earth is the water planet and about 97% of its water is in the oceans. About 2% is locked up in the ice caps and essentially removed from man's activities. That leaves about 1% in lakes, rivers, and the atmosphere for man's use. Although this seems like a small amount it is plenty for all of man's needs and, what's even better, that 1% is constantly being recycled through the system (see The Hydrologic Cycle, Chapter 9). Table 12-1 shows the estimated distribution

TABLE 12-1
ESTIMATED DISTRIBUTION OF WATER ON EARTH

Location	Volume (miles^3)*	Percent Total
Surface Water		
Lakes	30,000	0.009
Salt Lakes & Inland Seas	25,000	0.008
Stored in streams	300	0.0001
Subsurface Water		
Water in unsaturated zone	16,000	0.005
Groundwater to 2500 feet	1,000,000	0.31
Deep groundwater	1,000,000	0.31
Other		
Ice caps and glaciers	7,000,000	2.15
Atmosphere	3,100	0.001
World Ocean	317,000,000	97.2
Total (rounded)	**326,000,000**	**100.0**

*A cubic mile of water equals 1.1 trillion (1,100,000,000) gallons! After Feth, J. H., 1973, Water Facts and Figures for Planners and Managers, U.S. Geol. Survey Circular 601-I.

of water on Earth. Notice the small percentage of water in the atmosphere. It is the Earth's atmosphere, constantly in motion, that brings evaporated water and heat from the oceans to the land masses for use by man.

WATER USE

Water use can be broadly divided into three categories: domestic (family use), industrial (business and industry), and agricultural. Home use is fairly low and ranges from 10 gallons per capita per day (gpcd) to 80 gpcd. Highest use is found in areas served by electricity, and the affluent are larger consumers of water than low- and middle-income families. This is how we use water in our homes:

Toilet flush	3 gallons
Tub bath	30–40 gallons
Shower	20–30 gallons (unless a teenager)
Wash dishes	10 gallons
Washing machine load	20–30 gallons
Waste	
Dripping faucet	1 drip per second = 1,460 gallons per year
Humming toilet	1½ gallons/hour = 13,000 gallons per year
Watering Lawn	
Arid climate	36 in. per year on 8,000 square feet = 180,000 gallons
Humid climate	6 in. per year on 8,000 square feet = 30,000 gallons

In 1960, business and industrial use averaged 70 gpcd (Feth, J. H., 1973, Water Facts and Figures for Planners and Managers, U.S. Geol. Survey Circular 601-I). Add 10 gpcd public use for firefighting, street cleaning, park maintenance, and the like. The total of domestic, commercial, and public per capita use of water in American cities amounted to about 150 gpcd. This excludes agricultural use, whose figures are truly staggering.

By 1970, per capita use of water in the U.S. exclusive of agriculture was 180 gpcd. Of this, about 75–100 gpcd was discharged through sewers. Figure 12-1 shows how water was consumed in Kansas City, Missouri, in 1954. Figure 12-2 illustrates per capita water use in 100 of America's largest cities in 1962. The median per capita consumption was 140 gallons per day per person.

MEASURING WATER

Probably no substance has more varied units for measuring volume and flow than does water. Flow can be measured in gallons per minute (gpm), cubic feet per second (cfs), cubic meters per second (m^3/s), or the obsolete and easily forgotten "miner's inch." Volume on the other hand can be measured by the acre-foot (ac-ft), gallon, cubic foot (as on your water bill), or cubic meter. To give some order to these varied units let's organize them as follows:

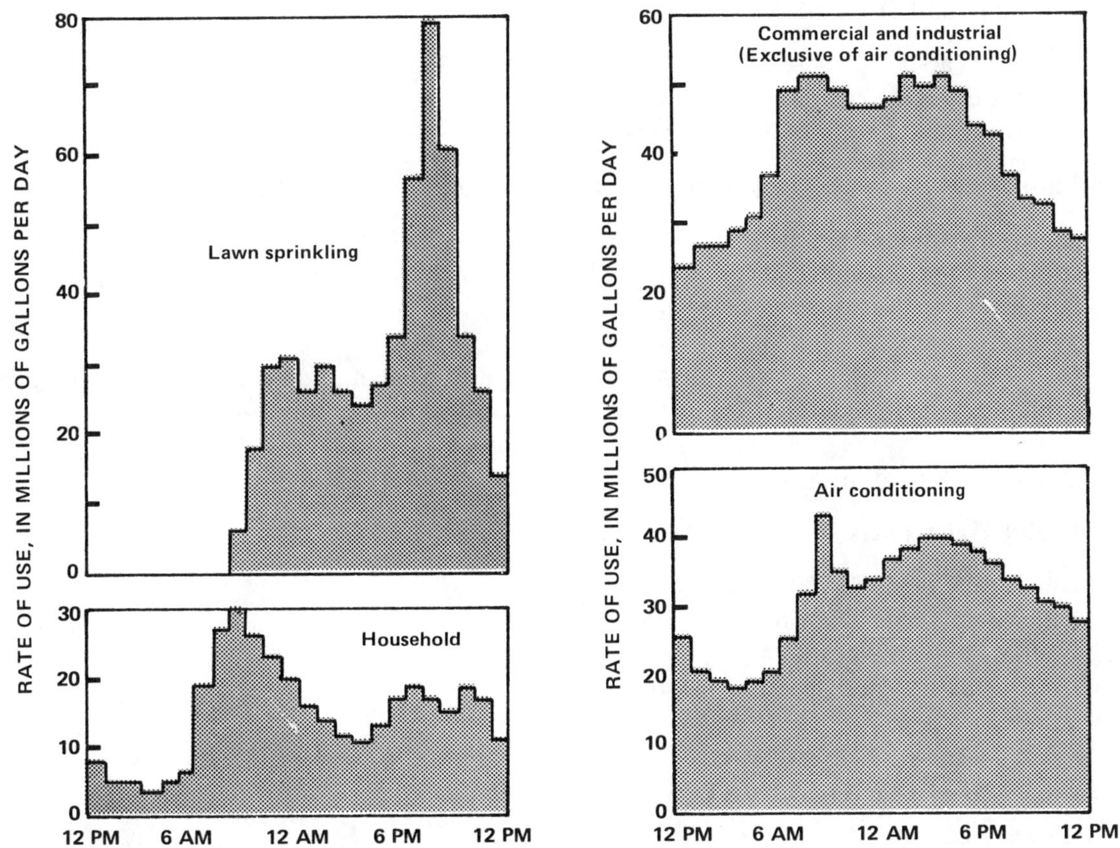

FIGURE 12-1. Hourly trends in water use on day of maximum use in 1954 (July 12), Kansas City, Missouri (Feth, J. H., 1973, Water Facts and Figures for Planners and Managers, U.S. Geol. Survey Circular 601-I).

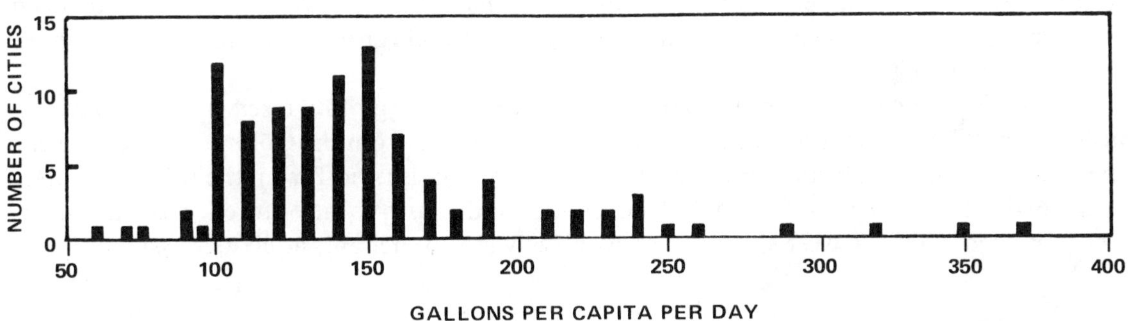

FIGURE 12-2. Per capita use of water in 100 largest cities of the United States in 1962. (After Feth, J. H., 1973, Water Facts and Figures for Planners and Managers, U.S. Geol. Survey Circular 601-I.)

Irrigation and Domestic Use

acre-foot (ac-ft) = 1 acre of water 1 foot deep = 43,560 cubic feet
cubic foot (cf) = 7.48 gallons
acre-foot (ac-ft) = about 326,000 gallons

Flow Rates (streams, pipes, springs, ocean currents)

gallons per minute (gpm)	=	1,440 gallons per day (gpd)
cubic feet per second (cfs)	=	7.48 gallons per second past a point in a stream or aqueduct
	=	449 gpm
	=	646,000 gpd or 0.646 million gallons per day (mgpd)
cubic meter per second (m^3/s)	=	22.8 million gallons per day (gpd)
	=	15,800 gpm
miner's inch (California)	=	1/30 cfs

Useful Water Facts

1 inch rain yields 27,200 gallons per acre
1 inch rain yields 100 tons water per acre
1 gallon water weighs 8.34 pounds
1 cfs yields 1.85 acre-feet/day (ac-ft/d)

WATER IN SEMI-ARID CLIMATES

In semi-arid and arid regions of the United States, local precipitation and surface water are insufficient to supply the water needs of populated areas. To satisfy these needs, water is imported by aqueduct from areas where there is an excess, and underground reservoirs are exploited. Southern California is perhaps the best example of large-scale management of water resources (Colorado east of the Rockies is similar). The average rainfall in California is 23 in./y (58 cm/y). About 30% reaches the ocean as surface runoff, some infiltrates ground water supplies, and the biggest percentage is lost to evaporation and transpiration (evapotranspiration). Although runoff exceeds the annual demand throughout California by a factor of three, a major problem exists: rainfall is variable and areas with the greatest demand for water do not have the highest average annual precipitation.

Because demand exceeds the total of local storage in underground reservoirs plus precipitation, southern California must import great volumes of water from the Colorado River, the east slope of the Sierra Nevada, and Oroville dam, which impounds runoff from the west slope of the Sierra north of Sacramento. Similarly Denver, Colorado, uses locally stored water as well as water imported from the west slope of the Rockies via a tunnel and aqueduct system.

EXERCISE XII

WATER RESOURCES

Name ______________________________

1. A common device for portraying rainfall on a map uses **isohyetals**. These are contours joining all points of equal precipitation, usually 1-, 5-, or 10-inch intervals. Thus all the area between the 10- and 20-inch isohyetal lines receives between 10 and 20 in. of rainfall per year (Fig. 12-3). Study the rainfall map and determine which areas receive the highest rainfall. Why? ______________________________

____

____

____

____

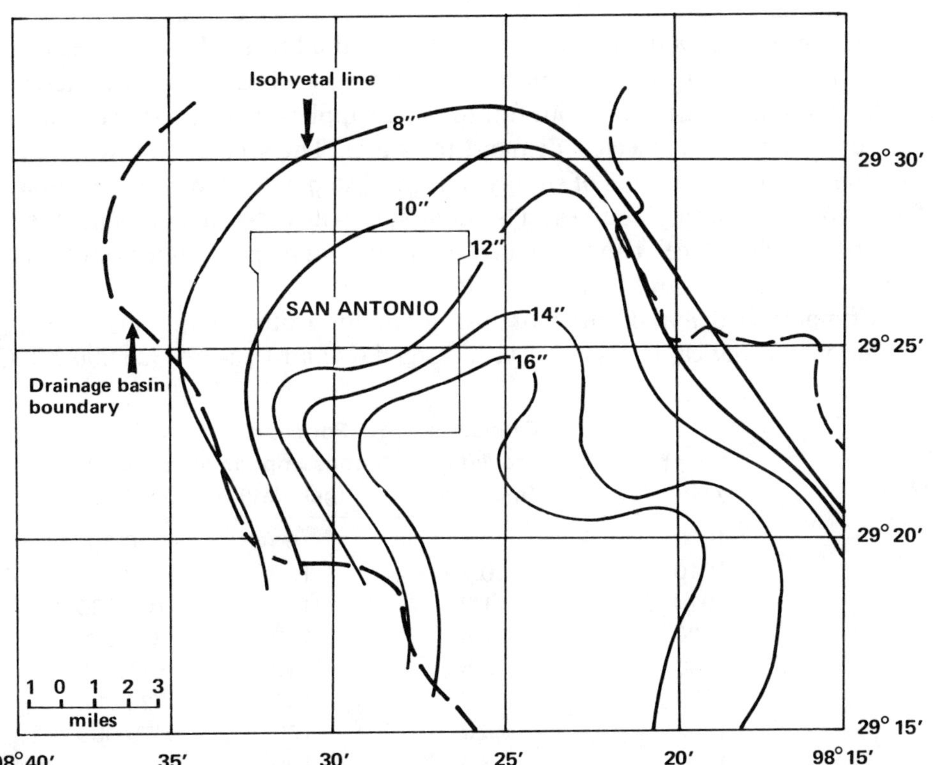

FIGURE 12-3. Isohyetal map showing distribution of rainfall on the land. (Rantz, S.E., 1971, Basic Data Contribution 25, U.S. Geological Survey.)

2. One unit often used to measure water volume is the acre-foot (ac-ft). This is the amount of water required to cover one acre to a depth of one foot and is equal to about 325,000 gallons. The commercial uses of water in California by sectors of the economy (1965 statistics) are:

		Acre-feet per year	% of total
Municipal		2.9×10^6	8.5
Industrial		1.3×10^6	3.8
Agricultural		29.8×10^6	88.7
	Total	34.0×10^6	100.0

Here is a philosophical question. If you were a commissioner on the Water Resources Board of California, how would you cut back water allocations during a drought that reduces supplies to 50% of normal levels? ______________________________

____

____

____

3. Imagine the year is 1900 and you have just been hired by the Los Angeles Department of Power and Water. The population of the Los Angeles Basin has been increasing rapidly and decisions must be made about future water supplies. The first step in your analysis will be to determine the projected **demand** for water. The second step will be to assess the **potential** for various sources of supply. The third step will be to do a **cost analysis** for each of the potential supply sources. The final step will be to recommend the priority for developing each source. The following questions will help to guide your thinking and are simplifed approximations to the real situation.

Compute a demand curve or histogram for water in this area, using estimated population and per capita usage. 1 year = 365 days and 1 ac-ft ≈ 326,000 gal. Check your units.

year	Estimated population (millions)	Per capita consumption (gal/day)	Annual demand (acre-feet)
1900	0.3	90	30,300
1910	0.7	100	78,600
1920	1.3	110	161,000
1930	2.9	120	391,000
1940	3.5	130	511,000
1950	4.7	140	723,000
1960	6.0	160	1,078,000
1980	8.0	180	
2000	11.0	200	

Step 1. Plot demand (ac-ft/yr) vs. time from 1900–2000.

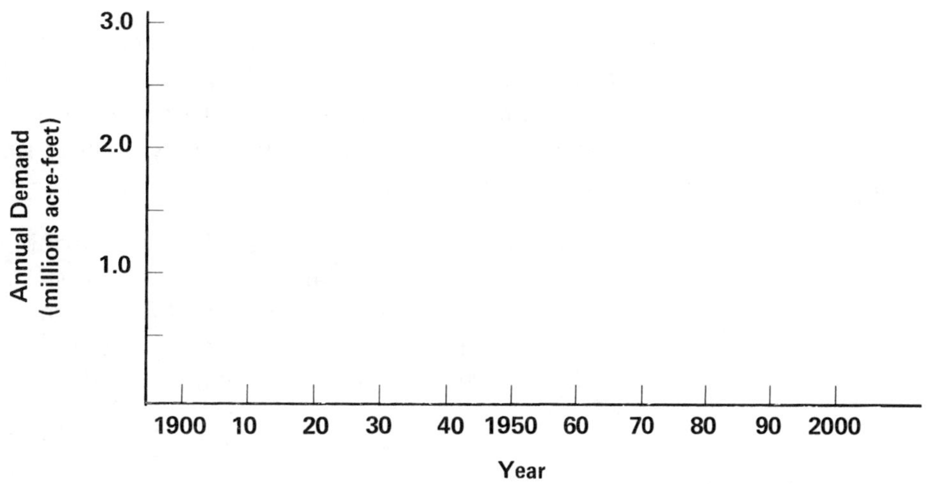

Step 2. Each of the following areas is a potential source of municipal water. The annual yield of each is summarized in the chart and detailed below (a through e). Calculate the contribution of the Sacramento River to the cumulative annual yield and fill in the chart.

Source	Annual yield for municipal use (ac-ft)	Cumulative yield (ac-ft)
(a) Groundwater pumping and the L.A. River	0.16×10^6	0.16×10^6
(b) Owens River Basin	0.5×10^6	0.66×10^6
(c) Mono Basin	0.3×10^6	0.96×10^6
(d) Colorado River Basin	1.2×10^6	2.16×10^6
(e) Sacramento River Basin		

(a) The safe annual yield from the L.A. River and from pumping groundwater is about 1.6×10^6 ac-ft. Only about 10% of this is of good enough quality for municipal water supplies. The remainder is used for irrigation.

(b) The average annual flow from the Owens River (excluding the Mono Basin) will yield 0.4 x 10^6 ac-ft. By drilling wells and pumping in this basin, an additional 0.1 x 10^6 ac-ft can be obtained.

(c) By diverting streams to flow into the Owens River from Mono Lake, an additional 0.3 x 10^6 ac-ft can be obtained. Mono Lake is a closed basin lake (no outlet) and loses water only by evaporation. Its level will fall at 1.6 ft/yr if these streams are diverted.

(d) The Colorado River is of marginal value for drinking water because of its high salinity and may be best suited for irrigation. Several other states and Mexico would like to utilize this resource. Of the water entering the lower Colorado basin, the following allocations can be made:

	Annual flow (10^6 ac-ft)
Average flow into lower Colo. River	8.2
Arizona	1.1
Nevada	0.004
California	5.1
Mexico	1.5

Of the water allocated to California, 1.2 x 10^6 ac-ft will be available for municipal supplies, with the remainder used for irrigation.

(e) The California Water Plan will be designed to transport water from the Sacramento River drainage basin to southern California. The maximum yield is 13 x 10^6 ac-ft per year, but it is unlikely that this yield can be attained because of agricultural demands and the cost of pumping.

Step 3. Making an accurate cost estimate for water supply is difficult. Most of the cost is for construction of an aqueduct and pumping. Assume that the cost of one ac-ft will be about \$0.10 per mi for aqueduct construction and that it takes one kwh/ft to pump water uphill at a cost of \$0.02/kwh. Multiply this figure by a factor of 1.5 to get the retail price of water.

Example: Water from Owens valley must be transported about 230 miles. To reach L.A. this water must be pumped about 500 feet uphill. The delivery cost will be about

1.5 (230 mi x \$0.10/mi) + 1.5 (500 ft x \$0.02/ft) = \$49.50

Now fill in the following table.

Source	Cost of aqueduct		Cost of pumping		Total cost per ac-ft
(a) Local (LA River etc.)	1.5 (10 mi x $.10)	+	1.5 (250 ft x $.02)	=	$ 9.00
(b) Owens River Basin	230 mi		500 ft		$49.50
(c) Mono Basin*	60 mi				$58.50
(d) Colorado River Basin	260 mi		500 ft		
(e) Sacramento River Basin	400 mi		2500 ft		

*Add cost of transportation of (c) to that of (b). Mono Basin is just north of Owens River.

Now plot the incremental cost-supply curve (acre-feet versus cost of additional acre-feet).

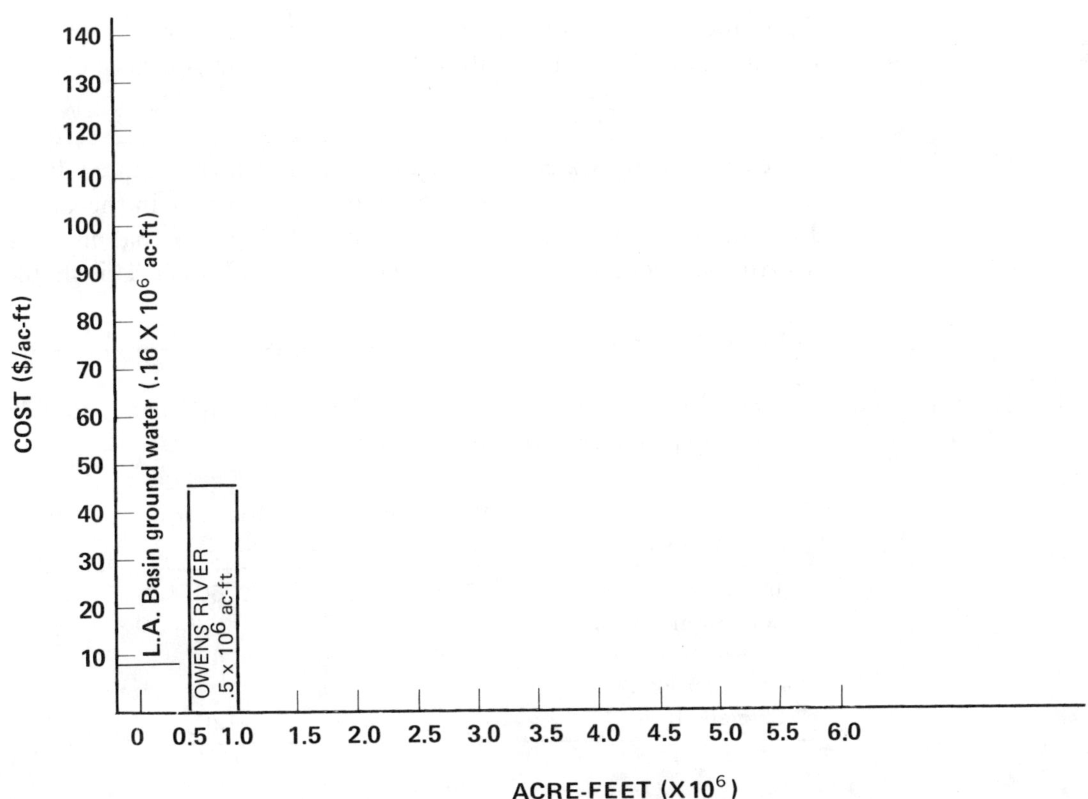

Step 4. Now that you have done the arithmetic, make some recommendations.

(a) In what year will each of the sources outside the L.A. Basin need to be tapped? (See graph of Step 1.)

Source	Cumulative yield (ac-ft × 10^6)	Year needed
Owens River	less than 0.16	1920
Mono Basin	less than 0.66	1945
Colorado River	less than 0.96	________
Calif. Water Plan (Sacramento River)	less than 2.16	________

(b) What was the average cost of water in 1980 and what was the incremental cost (the cost of one additional ac-ft) of water in 1980?

Average cost per ac-ft ____________________.

Incremental cost per ac-ft ____________________.

Think about the significance of this in allocating resources. Would you suggest that additional irrigation be encouraged in southern California?

____

(c) For extra credit: Calculate the amount of water required from the California Water Plan to meet the demand for water in the year 2000. How many kilowatt hours will this require? (Note that in 1975 the Los Angeles Dept. of Power & Water sold 17 x 10^9 kwh. This should put your calculation in perspective.) ______________________________

4. The projections you made in Problem 3 are based on data collected during the 1970s. In 1981 the actual cost of water delivered to Los Angeles was:

Source	Percent of total water	Cost per ac-ft
Ground water pumping	15	$ 60
Owens River/Mono Basin combined	80	$ 40
Calif. Water Plan/ Colorado River	5	$120

(a) Why do you suppose the cost of ground water increased so greatly over our earlier projection?

____

____

(b) What is the average cost of water delivered to Los Angeles from all sources in 1981? You must use the weighted average from each source, that is, percent of total water times the cost. For example, the cost of that part of an ac-ft of water that comes from ground water is 0.15 x $60.00 = $9.00. Do the same for the other sources and add all the cost components to find the total cost.

Total cost is $ ______________ /ac-ft

(c) Using the same logic calculate the average cost of an ac-ft of water as projected.

Source	Percent of total water	Actual cost per ac-ft	Projected cost per ac-ft
Ground water pumping	7	$ 9.00	________
Owens River	23	$ 49.50	________
Mono Basin	14	$ 58.50	________
Colorado River	56	$ 54.00	________

(d) How do the actual and projected costs compare? Identify sources of error or unforeseen events that would cause the prediction to deviate from actual cost.

____

____

____

Chapter 13

Mineral Resources and Fossil Fuels

Civilization is dependent on extractable or usable mineral and energy resources. Every day the world consumes mineral products and fossil fuels. The amount consumed is a function of a society's standard of living, with the most affluent societies being the greatest consumers of energy and mineral products (Fig. 13-1). Unlike timber and water resources, minerals and most forms of energy are **not** renewable; once consumed they are gone forever. If we want to maximize the benefits to man of the Earth's mineral resources and fossil fuels, we must manage them wisely.

The quality of life may be expressed by a simple equation (McKelvey, 1973, "Mineral resource estimates and public policy," U.S. Geol. Survey Circular 682):

$$L = \frac{R \times E \times I}{P}$$

where L is the standard of living measured by consumption of goods and services, R is raw materials consumed (metals, nonmetals, water, soil, wood fiber, and so on), E is consumption of

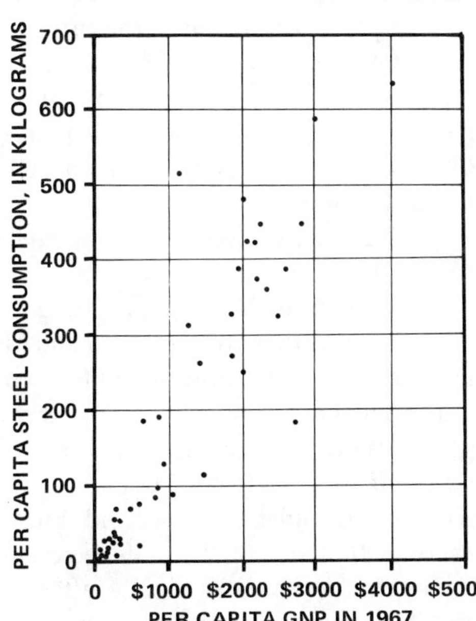

FIGURE 13-1. Per capita steel consumption compared to per capita Gross National Product (GNP) for countries that appeared in the U.N. Statistical Yearbook, 1967.

all forms of energy, I is "ingenuity" (political, technical, and socioeconomic), and P is the number of people who share the resources. Raw materials are a key element in this equation, for without them there cannot be a high standard of living. If large amounts of raw materials, energy, and ingenuity are shared by a relatively few people, a high material quality of life results. The factor most difficult to evaluate is ingenuity. A highly imaginative society can live better on fewer resources than one that is short-sighted and indulgent.

CLASSIFICATION OF RESOURCES

A mineral resource is a concentration of elements in rocks or in the ocean in such a form that a usable commodity can be extracted from it. The commodity can be an element, such as iron or aluminum; a compound, such as salt or borax; a mineral, such as talc or asbestos; or simply a useful rock, such as limestone or gypsum. Minerals can be subdivided further into **metallic** and **nonmetallic** ones. Metallic minerals can be mined for their metal (e.g., copper); nonmetallic industrial minerals are extracted for various uses from plaster to concrete aggregate.

Fuels such as oil, gas, and coal present a special case because, strictly speaking, they are not minerals. They are commonly called "fossil fuels" because they represent stored-up solar energy in the form of plant or animal remains, or simply "mineral fuels." Fossil fuels make the U.S. economy go and lack of them is reflected in trade deficits abroad and inflation at home.

FOSSIL FUELS

The industrial revolution of the mid-19th century was triggered by development of coal-based technology. In 1859 the first oil well was drilled in Pennsylvania and eventually spawned the largest private enterprise on Earth. About half the geological profession is employed in the search for oil and other fossil fuels. Much of this chapter is devoted to an analysis of our fuel resources, but keep in mind the principles used are also applicable to "hard" minerals.

Petroleum is a mixture of hydrocarbons (carbon-hydrogen compounds) that occur as liquid, solid, or gas. The compounds range from naturally occurring methane (CH_4), called natural gas, to very complex distillates like benzene and naphtha, which combine to form crude oil and which are fractionated, cracked, or distilled in oil refineries to produce the various forms useful to man (Fig. 13-2).

There are four conditions for the formation and accumulation of oil in nature: (1) a source rock, (2) a reservoir rock, (3) a cap rock, and (4) a suitable structure. **Source rocks** are usually sedimentary rocks high in organic matter. This organic material is largely the remains of microscopic marine plants that are preserved and altered in time to crude oil. Source rocks are generally fine-grained, e.g., shale, and the oil formed in them migrates into **reservoir rocks.** These are coarse-grained sedimentary rocks, such as sandstones or limestones, which have a high porosity and permeability (see Chapter 9). Porosities on the order of 40–50% mean that each cubic ft of reservoir rock has 0.4–0.5 cubic ft of oil in place. However, only 20–30% of the oil in pore spaces is recoverable under the best conditions. So that oil does not seep to the surface of the ground or become diluted in other rocks there must be an impermeable **cap rock** to contain the fluid. Shale is a common cap rock. Finally, after all the conditions for accumulation are met, there must be a **suitable structure** or trap to contain the oil. Image a teacup turned upside down into a pan of

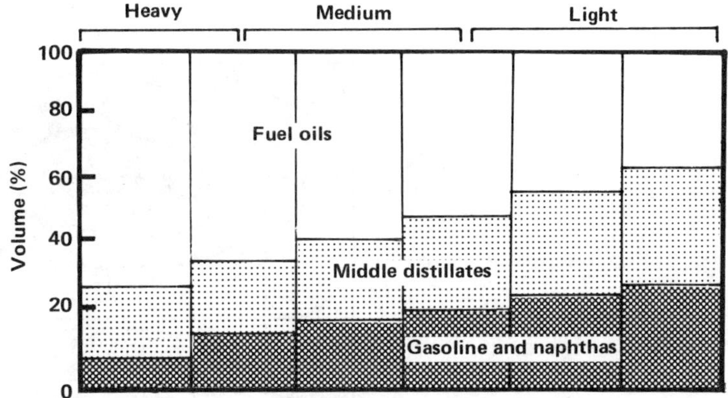

FIGURE 13-2. Products of refining six different crude oils of varying quality.

water and you can picture an analog of an anticline trap for oil. Inside the teacup there is air and fluid, whereas the fluids trapped in an anticline (upfold) are gas, oil, and water. Several kinds of oil traps are shown in Figure 13-3. With ingenuity and skill, the petroleum geologist can locate these traps in oil-producing provinces. Of course, the ultimate test is the costly drilling of a well to see if the geologic interpretations were correct and the strata are indeed oil-bearing.

RESOURCE TERMS DEFINED

Precise definitions of resource terms are necessary to understand the present and future needs of our resource base, in particular, oil and gas. The following terms are used in this chapter:

Reserves. That portion of the identified resource which can be economically extracted.

Measured reserves. That part of the identified resource which can be economically extracted using existing technology and whose amount is supported by geologic engineering evidence.

Indicated reserves. Reserves that can only be extracted by secondary recovery techniques, such as fluid injection. These reserves are **in addition** to the measured reserves in a known reservoir.

Demonstrated reserves. A collective term for the sum of measured and indicated reserves.

Inferred reserves. Reserves in addition to demonstrated reserves eventually to be added to known oil and gas fields through extensions, revisions, and new producing zones.

Cumulative production. The sum of production for the current year and the actual production for all prior years.

Undiscovered resources. Those economic resources which are **estimated** to exist in favorable geologic settings but are yet undiscovered.

Several points are worthy of note. Inherent in the concept of "reserves" is the ability to exploit them economically or, if you wish, at a profit. Most of the statistics used in this chapter are for 1974. Oil prices have risen enormously since that time. This rise in dollar value has increased the demonstrated reserves as many previously uneconomic oil accumulations became economically attractive.

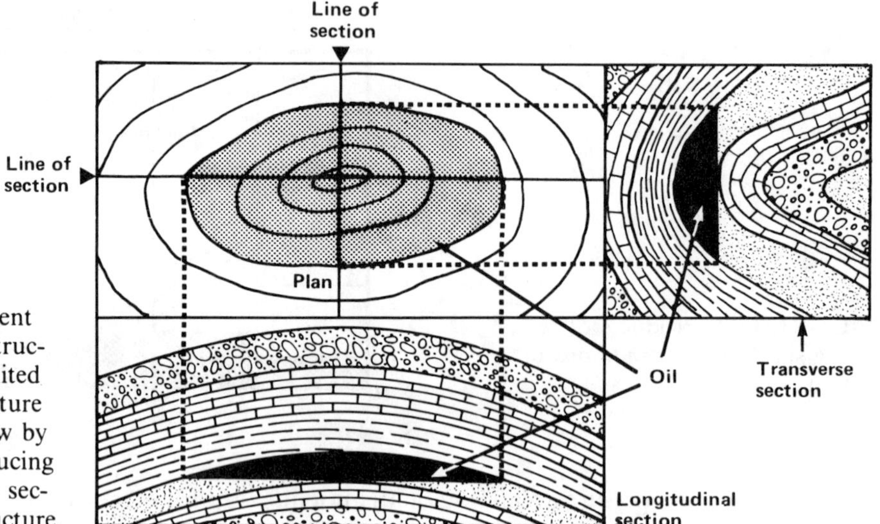

FIGURE 13-3. Different kinds of oil-bearing structures found in the United States. a. Dome structure illustrated in plan view by contours on the producing reservoir sand and by sections through the structure.

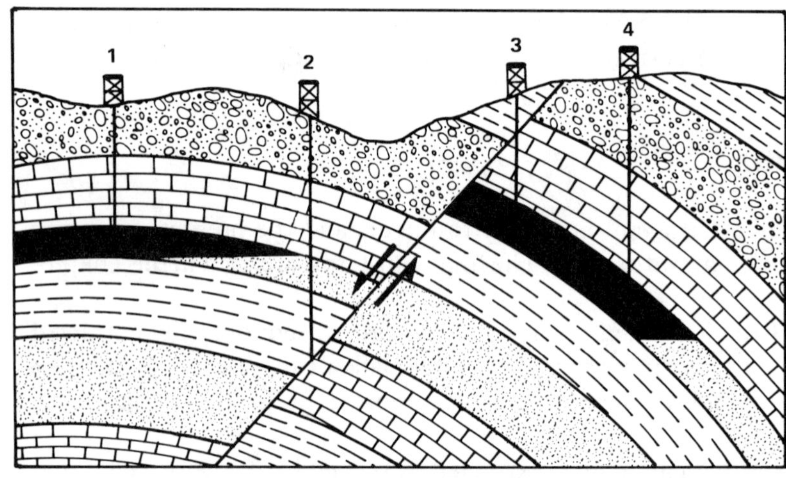

b. A faulted anticline showing how faulting may leave barren zones in the structure (Well 2). Note that Well 3 was drilled through the fault zone.

Limestone
Anhydrite
Salt dome

c. A salt dome typical of the Gulf Coast with oil trapped against the flank of the dome and in porous limestone along the crest. (After Uren, L.C., 1939, Petroleum Production Engineering, New York: McGraw-Hill Book Co.)

TABLE 13-1
TOTAL KNOWN AND INFERRED OIL RESERVES IN THE UNITED STATES TO DECEMBER 31, 1974 (Billions bbl)

	Cumulative Production	Demonstrated Reserves Measured	Demonstrated Reserves Indicated	Inferred Reserves	Undis-covered (mean)
Onshore					
Total lower 48	99.9	21.1	4.3	14.3	44
Total U.S.	100.0	31.0	4.3	20.4	56
Offshore					
Total lower 48	5.6	3.1	0.3	2.6	11
Total U.S.	6.1	0.3	0.3	2.7	26
Total U.S. off- and onshore	106.1	34.2	4.6	23.1	82

RESERVES

Petroleum. Estimates of reserves and undiscovered reserves of crude oil are shown in Table 13-1 and Figure 13-4. These data are best estimates based on extensive geologic and statistical analyses by a team of scientists from government agencies. The economics inherent in the definition of "reserves" makes the 82 billion bbl of undiscovered oil and the 120–140 billion bbl of sub-economic oil very promising indeed. The same analysis has been applied to natural gas (Fig. 13-5). There are 237 trillion cubic feet (TCF) of known reserves and 300 TCF of inferred and indicated reserves.

The bottom line of Table 13-1 indicates about 39 billion bbl of demonstrated reserves and about 105 billion bbl of undiscovered and inferred reserves. At high current prices, inferred and indicated reserves become very attractive. The same applies to our natural gas resources.

Coal. The U.S. has the largest coal reserves on Earth, about one-fourth of the 786 billion tons known to exist in the world. Coal constitutes 80% of our fossil fuel reserves but only 18% of the demand. Most of the coal used is converted to electrical energy in steam-powered generating stations. There is enough coal in this country to last at least 200 years at present rates of consumption; however, problems plague this old fuel. One of the most serious obstacles at present is emissions of sulfur from coal-burning sources that produce "acid rain" and corrode buildings, render lakes sterile, and disturb human health.

Tar Sand. A fossil fuel source of great promise is tar sand. In this resource the hydrocarbon is very thick and must be "washed" from its host rock. It is estimated there are 300 billion bbl of oil tied up in tar sands and that 50,000 bbl/day could be extracted now from deposits in Utah. The most promising reserves are in California, Texas, and Utah.

Oil Shale. Oil shale is another promising source of energy. The hydrocarbon (kerogen) in oil shale is in solid form in the rock. The shaley rock must be broken and heated to release the fluid. The amount of oil obtained ranges from 15 to 40 gallons per ton of rock and it is estimated that no less than 2 trillion bbl of oil are tied up in shales in the western United States. The hope is to extract the oil in place so that disruption of the surface land is mimimized.

	IDENTIFIED RESOURCES			UNDISCOVERED RESOURCES
	DEMONSTRATED			
	Measured	Indicated	Inferred	
	RESERVES			
ECONOMIC	34.250	4.636	23.1	50–127
SUB-ECONOMIC	120–140			44–11

◀ Increasing degree of geological assurance

Increasing degree of economic feasibility ▶

Total U.S. Cumulative Oil Production 106 billion barrels 12/13/74

FIGURE 13-4. Crude oil resources of the U.S. in billions of barrels. (From Miller, Betty M., and others, 1975, "Geological estimates of undiscovered recoverable oil and gas resources in the United States," U.S. Geol. Survey Circular 725.)

	IDENTIFIED RESOURCES		UNDISCOVERED RESOURCES
	DEMONSTRATED		
	Measured	Inferred	
	RESERVES		
ECONOMIC	237.132	201.6	322–655
SUB-ECONOMIC	90–115		40–82

◀ Increasing degree of geological assurance

Increasing degree of economic feasibility ▶

Total U.S. Cumulative Gas Production 481 trillion cu ft 12/31/74

FIGURE 13-5. Natural gas resources of the U.S. in trillions of cubic feet. (From Miller, Betty M., and others, 1975, "Geological estimates of undiscovered recoverable oil and gas resources in the United States," U.S. Geol. Survey Circular 725.)

TABLE 13-2
ENERGY UNITS AND EQUIVALENTS

BRITISH THERMAL UNIT (Btu) = amount of heat required to raise the temperature of 1 pound of water 1°F
QUAD = one quadrillion Btu = 172 million barrels of oil
BARREL OF OIL (bbl) = 5.8×10^6 Btu = 42 gallons
NATURAL GAS = variable Btu content; reserves in trillion cubic feet (TCF)

ENERGY CONSUMPTION

The amount of fuel energy consumed in the United States is shown in Figure 13-6. A "quad" is a quadrillion British thermal units (Btu) and is equivalent to approximately 172 million bbl of oil (Table 13-2). A futuristic view of our energy consumption is shown in Figure 13-7. It seems inevitable that energy demands will increase worldwide as well as in the United States. Total world energy use is projected to increase about 56% during the 15 years from 1975 to 1990, and the demand for crude oil will likely increase 58% in the same period. This projection may not hold for the U.S. if government energy policy stimulates development of solar, geothermal, wind, and synfuel alternatives. (Synfuels include such sources as gasification of coal, oil shales, and gasohol.)

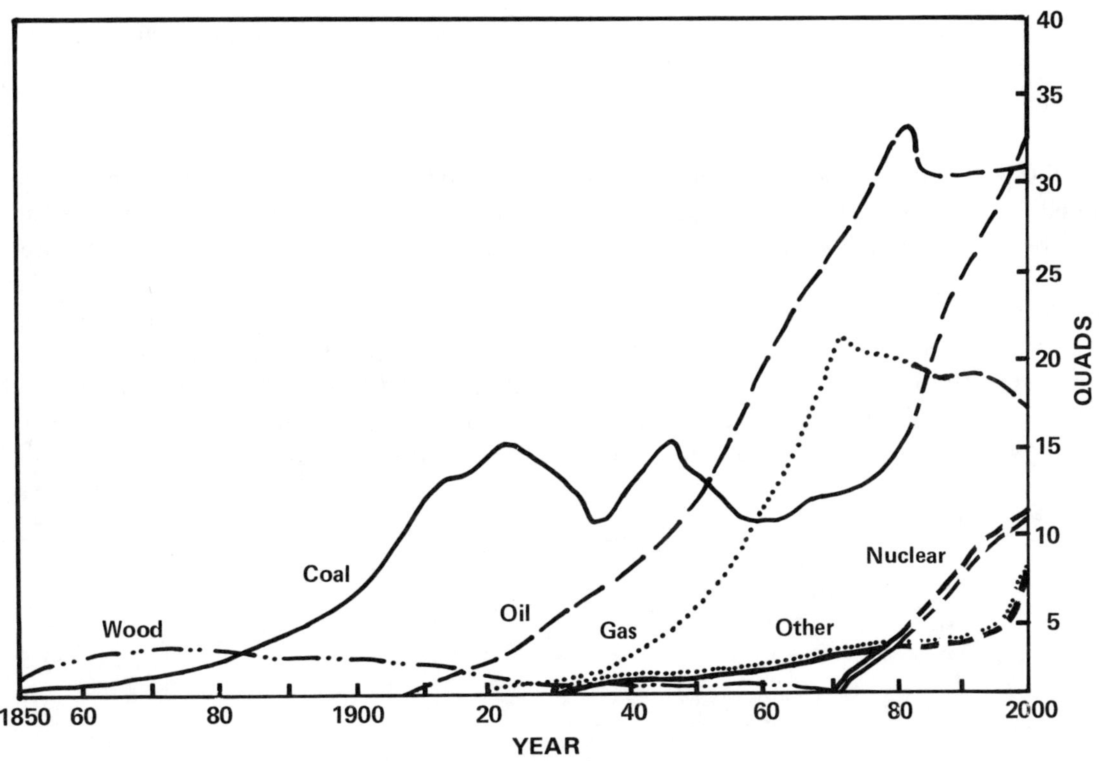

FIGURE 13-6. Allocation of energy usage in the United States. After year 1980, curves are estimates. (Energy Information Administration.)

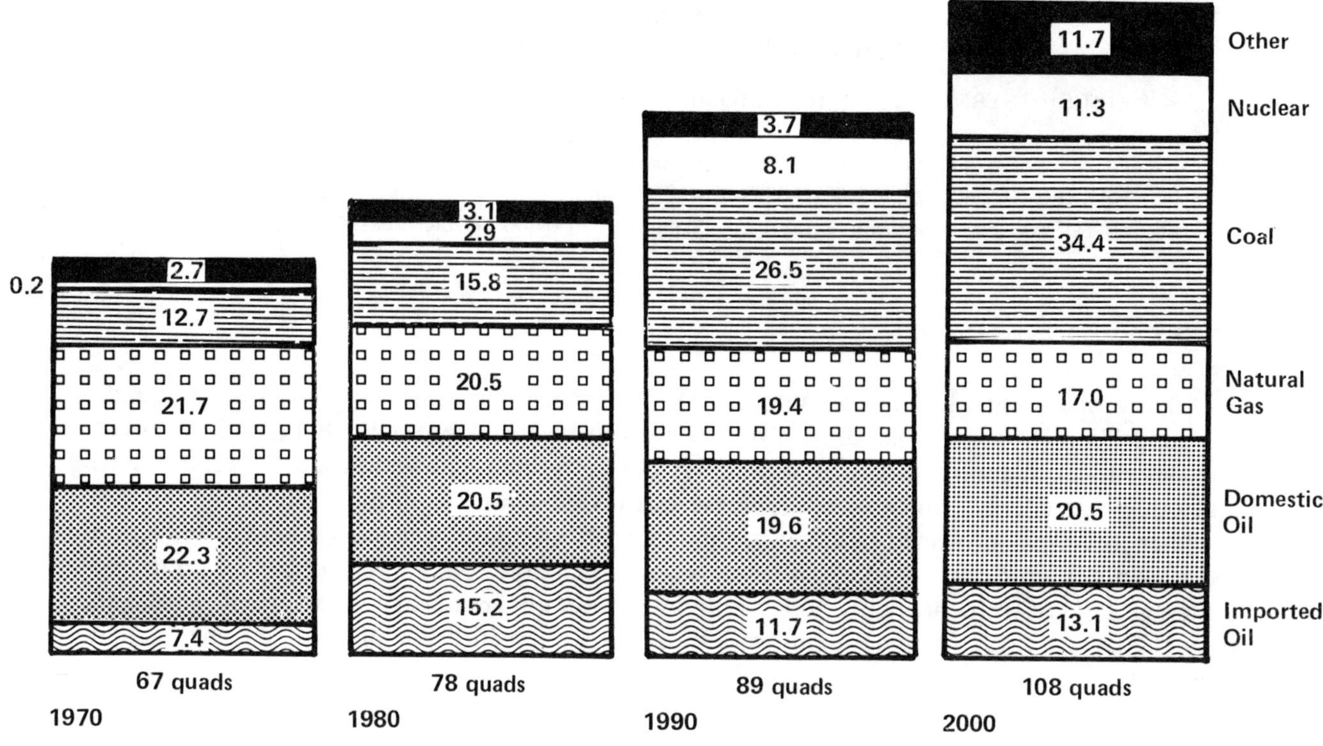

FIGURE 13-7. Total energy consumption and allocation in the United States and projections to the year 2000 (Energy Information Administration).

Current world consumption of crude oil is 56 million bbl/day or about 20 billion bbl/year. United States consumption is 10 million bbl/day domestic and 7 million bbl/day imported crude oil. The U.S. uses 6.2 billion bbl/year of crude oil—almost one-third of the consumption worldwide.

Natural gas is produced from an estimated 165,000 wells across the country and supplies 26% of our energy needs. Production in 1979 exceeded discoveries by 39% and our reserves dwindled to 195 TCF. In 1980 we consumed about 20 TCF or 10% of our reserves without further discoveries. The desire to import liquefied natural gas (LNG) from foreign countries and from Alaska was based on such production and use considerations.

EXERCISE XIII

MINERAL RESOURCES AND FOSSIL FUELS

Name ______________________________

1. By using Figures 13-6 and 13-7 determine the change in pattern of energy use from 1980 to the year 2000.

   Domestic oil ______________________________

   Imported oil ______________________________

   Coal ______________________________

   Nuclear power ______________________________
   What are the sources of energy called "other" in the two figures?

   ____

2. From Figure 13-4, determine how many years the demonstrated U.S. reserves of crude oil will last at the current rate of consumption without imported oil. ______________

   How long will they last at the current rate with imported oil? ______________
   Assume that identified sub-economic oil deposits become economical. What lifetime will this add to U.S. energy independence at the current rate of consumption? ____________
3. How many years will the demonstated U.S. reserves of natural gas last at the current rate of consumption (from Fig. 13-5)? ______________________________
4. Since 1978 the U.S. government has gradually decontrolled the price of natural gas. What will be the effect of this relatively new policy on natural gas reserves (examine Fig. 13-5)? ______________________________
   What will be the effect of decontrol on the consumer?

   ____

5. What must be done to relieve U.S. dependence on crude oil and particularly the dependence on imported oil? ______________________________

   ____

   ____

   ____

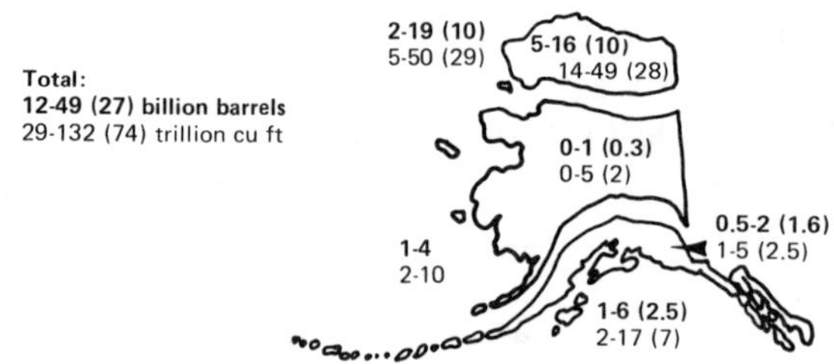

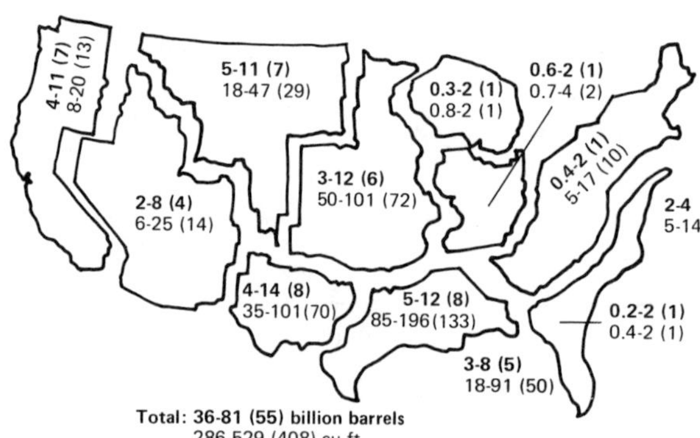

FIGURE 13-8. Undiscovered recoverable crude oil and natural gas onshore and offshore United States. Boldface numbers are for oil, lightface for natural gas. Numbers in parentheses are the estimated average reserves. In the range of values, we are 95% sure of the low figures and only 5% sure of the high. (Adapted from Miller, Betty M., and others, 1975, "Geological estimates of undiscovered recoverable oil and gas resources in the United States," U.S. Geological Survey Circular 725.)

6. On Figure 13-8, put a check mark by the three provinces with the greatest potential for "new" oil. Name one state or region within each of the three provinces. ____________

____

7. (a.) Locate and number the three best sites for exploratory oil wells on the structure contour map below. Contours are drawn on top of a possible oil sand.

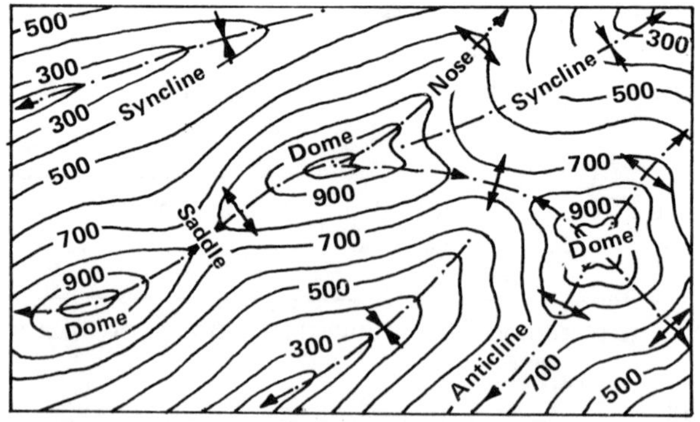

(b). The contours are elevations above sea level. If the ground surface were level and at elevation 5200 ft (Colorado, for instance), what would be the depth of each of the three wells you have located?

1. __________ feet. 2. __________ feet. 3. __________ feet.

Chapter 14

Pollution and Residence Time

Pollution is a subject much studied by environmental geology and includes pollution of air, surface water, and ground water. Air is polluted primarily from automobile exhaust, factory emissions, and volcanic eruptions. Surface waters (streams, lakes, bays, oceans) become polluted when noxious substances reach them via runoff or through the dumping of waste products into them. Ground water gets polluted when noxious substances like pesticides are leached from the surface of the ground to infiltrate subsurface water. Leaking septic tanks are another source of ground water pollution.

RESIDENCE TIME DEFINED

To study pollution systemically, we use the concept of **residence time,** which relates information about the rates at which pollutants flow into and out of water basins (like Lake Superior) or air basins (like the Los Angeles basin) and the amount of time the pollutants remain there. The basin may have well-defined boundaries, like Lake Superior, or ambiguous boundaries, like the Los Angeles basin. Most basins allow the pollutant to enter and leave, although not necessarily at the same rate.

We can measure the rates of inflow and outflow and thus determine how long a pollutant will remain in a basin. For convenience, we refer to the combination of pollutant and water or air as a **compound**. To measure residence time (R), we examine the ratio of amount of compound in a basin (C) to the instantaneous rates of inflow and outflow of the compound (F). This can be expressed by the equation

$$R = \frac{C}{F}$$

Residence time is valid only for conditions at the time of measurement.

SAMPLE PROBLEMS

Type I — Without Mixing. The simplest type of residence time problem is where the amount of compound leaving a basin equals the amount of compound entering the basin, and the incoming compound does not mix with the compound in the basin. Suppose a stream flows at 90 gal/min into a reservoir holding 9 million gallons and that the reservoir drains at 90 gal/min. Since we are assuming that mixing does not occur, we can think of the incoming water as physically displacing the water in the reservoir. To simplify the problem further, we assume that no water evaporates and that none is lost through infiltration to groundwater. The residence time of the water in this reservoir, then, is

$$R = \frac{C}{F} = \frac{9{,}000{,}000 \text{ gal}}{90 \text{ gal/min}} = 100{,}000 \text{ min} \approx 1{,}666 \text{ hr} \approx 70 \text{ days}$$

It would take 70 days from the time the pollutant enters the basin to the time it begins to leave.

Type II – With Mixing. A slightly more complex problem arises if we assume the waters do mix. At a specified time, the reservoir would contain a mixture of new stream water and old reservoir water. Using the above example, at the end of 70 days, the mixture of waters in the reservoir would be 50% stream water (WHY?). After another 70 days, the mixed water is again changed by 50% and the reservoir would contain 75% stream water and 25% old reservoir water (WHY?). After another 70 days, the reservoir contains 87.5% stream water. From this discussion we can see that reservoir water is replaced by new stream water at a constant ratio of one-half every 70 days. This constant ratio of replacement can be expressed mathematically in a convenient manner using an exponent. At the end of about 1 year (70 days x 5 = 350 days), the amount of original water remaining in the reservoir is $(\frac{1}{2})^5$, where ½ is the ratio and the exponent is the number of 70-day increments. By this analysis, the old reservoir water would never be completely flushed out (WHY?).

Suppose the water in a reservoir (6,000,000 gal) is draining faster than the water entering it. If the water drains at 120 gal/min, and the rate of inflow is 60 gal/min, how much water would be in the reservoir at the end of a specified time? At the end of, say, 30 days (about 43,200 min), the amount of water in the reservoir would be

$$6{,}000{,}000 \text{ gal} + (60 \text{ gal/min} \times 43{,}200 \text{ min}) - (120 \text{ gal/min} \times 43{,}200 \text{ min})$$

$$= 3{,}408{,}000 \text{ gal}$$

The residence time of the pollutant remaining in the reservoir after these 30 days can be approximated using the difference in the rates of flow (net outflow in this case) and the amount of water remaining in the reservoir:

$$R = \frac{3{,}408{,}000 \text{ gal}}{60 \text{ gal/min}} = 56{,}800 \text{ min} \approx 40 \text{ days}$$

The problems become even more complex if we have more than one source for adding materials and more than one source for subtracting materials. Water may be added to a reservoir from rain falling on the the reservoir; from several streams, each with a different rate of flow. Water may be subtracted by evaporation, by infiltration to the ground water, or by opening the reservoir gates. Even though the problems can become exceedingly complex, the principle for solving them remains the same.

EXERCISE XIV

POLLUTION AND RESIDENCE TIME

Name ______________________________

1. A lake contains 5 million gallons of water. A factory empties a liquid dye into a stream that flows into the lake.
   (a) The stream flows at 50 gal/min and the lake drains at 50 gal/min. How long will it take for the water containing the dye to replace the water in the lake? Assume the dye and the stream water do not mix with the water in the lake. __________
   (b) A different factory is located on the bank of the lake and a liquid pollutant has been allowed to enter the lake for a period of time so that the lake now contains 20% pollutant and 80% water. During this time, water did not enter the lake and the lake was not drained. Assume that the pollutant has completely mixed with the water in the lake. If the stream now begins to flow at 50 gal/min and the lake drains at 50 gal/min, and assuming the stream water does not mix with the polluted water in the lake, how long will it take for the amount of pollutant in the lake to be reduced to:

   10% (that is, one-half original amount)? ____________

   1%? ____________
2. Two streams, each having a different flow rate, enter a lake holding 3 million gallons of water. The lake is drained by another stream with a flow rate equal to the sum of the flow rates of the incoming streams. Each incoming stream contains a different pollutant: Stream A contains 20% liquid pesticide and Stream B contains 5% liquid dye.

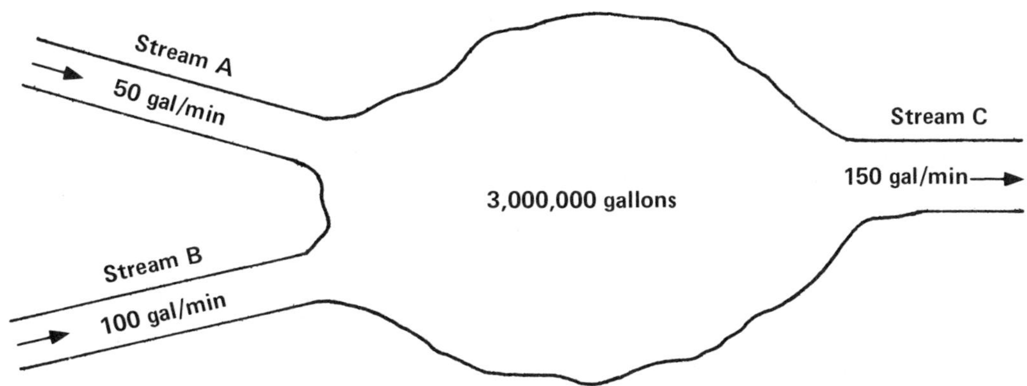

   (a) Assuming no mixing, how long would it take for one-half of the lake to become polluted? ______________________________

(b) In the polluted one-half, what percentage of water is pesticide? ________________

What percent is dye? ____

How many gallons of pesticide are in the lake? ___________________________

How many gallons of dye are in the lake? _______________________________

(c) When 87.5% of the original lake water has been displaced, what is the percentage of each pollutant in the lake? ____

How many gallons of each pollutant are in the lake? _______________________

(d) Suppose that a regulation has been passed prohibiting further pollution of the lake. If 87.5% of the original water in the lake had been polluted before the new law, how long would it take for pesticide A to be reduced to 1%? Assume that pesticide A has completely mixed with the water in the lake. ______________________________

Chapter 15

Environmental Impact Statements

by David Schleicher*

The primary purpose of an environmental impact statement (EIS) is to provide planners with information that aids them in making a decision about environmental projects. Corollary purposes are to inform the public of the bases for the decision and to encourage the public to review their validity. **The National Environmental Policy Act** of 1969 (**NEPA**) requires that an EIS be written for every major federal action that significantly affects the quality of the human environment. A "federal action" is typically either a federal proposal to do something (e.g., build a dam or lease lands for coal mining) or a federal review of somebody else's proposal (e.g., deciding whether to approve a mine plan proposed by a mining company). This chapter focuses on a federal review, specifically, consideration of a proposed plan for a coal mine.

A CASE IN POINT — A PROPOSED COAL MINE

An EIS for a proposed coal mine should answer two fundamental questions:

1. Should mining be allowed in the area under question? Generally, it is legally required to allow mining on lands leased expressly for that purpose. Mining may, however, be prohibited because of a hazard of monumental significance, such as the presence of endangered species on the leasehold. Such authorization to cancel a lease is rare.

2. If mining is allowed, which of the available alternatives should be selected — the mine as proposed? Or with added conditions? Or an entirely different plan? This is the more usual situation. Typically, a legal decision has already been made to allow mining. The question that remains is "According to what plan?" To answer this question we must be able to compare the hazards and benefits of the proposed plan with the hazards and benefits of alternatives.

DRAFT ENVIRONMENTAL STATEMENT (DES)

The first step in a formal review of environmental impact is the **draft environmental statement (DES),** which provides the public with information about the anticipated impacts of the

*U.S. Geological Survey, Denver, Colorado

proposal and its alternatives. It allows the public to verify whether the DES accurately depicts what would happen if the mine plan or one of its alternatives were approved and mining and reclamation were to take place accordingly.

The DES should make it as easy as possible for the public to check its conclusions about expectable impacts and their significance. Errors in the DES can be corrected for the **final environmental statement (FES)** and the decision-making summary derived from it. The decision maker and the general public must be able to understand the essential issues even if they do not fully understand the technical arguments that support those conclusions. Figure 15-1 shows a cover sheet for a typical DES.

Responsible agency: Office of Surface Mining Reclamation and Enforcement

Cooperating agencies: Interstate Commerce Commission
U.S. Bureau of Land Management
U.S. Forest Service
U.S. Geological Survey

Proposed action: Approval of the mining and reclamation plan submitted by Antelope Coal Company (NERCO) for the Antelope mine.

For further information: Florence Munter Schaller
Office of Surface Mining Reclamation and Enforcement
Brooks Towers
1020 15th Street
Denver, Colorado 80202

(303) 837-5656 (commercial)
327-5656 (FTS)

Type of statement: Draft environmental impact statement (EIS)

Abstract: Antelope Coal Company (NERCO) proposes to mine 260.3 million tons of coal over a period of 29 years. In the process, 5,860 acres would be disturbed. Three more coal mines have been proposed for the area. The Antelope mine, in conjunction with the other mines, would significantly impact cultural resources and create moderately significant socioeconomic impacts on the city of Douglas and Converse County, Wyoming. In addition, transportation impacts, especially from railroad traffic from the mine, would add significantly to the impacts already occurring in the region. Other impacts, which would be typical of surface coal mining operations in northeastern Wyoming, would be moderate or insignificant except that the cumulative removal of agricultural and wildlife lands would continue to increase until large areas were reclaimed and returned to agricultural and wildlife uses approximating premining levels. The Wyoming Department of Environmental Quality has identified deficiencies in the mine plan which must be corrected to bring it into compliance with Federal and State regulations.

Comments regarding this draft EIS must be received by: November 27, 1981.

FIGURE 15-1. A cover sheet for a typical draft environmental statement (DES).

ENVIRONMENTAL CONSEQUENCES (IMPACTS)

Environmental consequences or impacts of mining are the changes in the environment due to the mining. The postulated impacts are necessarily forecasts, and sometimes are no more than educated guesses. The discussion of impacts should identify not only the changes to the land but also the changes to everything else. For example, it should tell how much the population of a given town would increase because of the proposed mine, as well as how much the population of that town would increase without the proposed mine (and not necessarily in the absence of any other development). Similarly, it should tell how much the regional air quality would change with and without the proposed mine. In other words, the EIS really must forecast (generalized) impacts on everything in the impacted region.

Any discussion of impacts presupposes geographic boundaries for the impacted region. The boundaries are actually different for each discipline (the soils, wildlife, air, and so on) and even within a discipline (e.g., deer vs. field mice). Impacts for soils are pretty much confined to the mine. Those for wildlife depend on how far deer, let's say, are displaced from the disturbed area, how far they, in turn, displace other deer, and how much those displacements regionally impact habitat, carrying capacity, and herd size.

The discussion of impacts should consider the following question: "What aspect of each discipline can best sum up the impacts to be expected if mining and reclamation are done as proposed?" For soils, for example, we might ask, "What features of the soils best sum up the changes that the soils would undergo as a result of mining? Is it soil types? Horizons? Structures? Thicknesses?" Perhaps a better answer is "soil capabilities," that is, the uses the soils can support. Changes in the thicknesses of the soils are one measure of the **magnitude** of the impact, in crude terms, how big it is. Changes in the capabilities of the soils, by contrast, measure the **significance** of the impact, in crude terms, how bad it is. Such changes in the ability of each resource to support anticipated uses answer what is perhaps the fundamental question for the public reviewer and the decision maker, namely: "How concerned should I be about this impact? I know how big it is — but so what? What does it mean?" While we're certainly interested in how soil thicknesses would change, our real concern is whether the soil can support anticipated uses after reclamation. If they can, we're not especially concerned about long-lasting impacts; if they cannot, we must recognize this change as a significant environmental cost of mining.

Let's examine the distinction between magnitude (how big) and significance (how bad) in more detail.

Magnitude. The magnitude of an impact varies in space and time: it may be greater here than there, less then than now. The magnitude of the impact, together with its variation in space and time, answers the questions: "How big is the impact? How intense? How extensive? How long-lasting?" Magnitude is expressed, for example, as the amount of change in elevation and slope of the land surface, in air and water quality (concentrations), in wildlife population, in human population, in budget deficit. Ideally, it is possible to draw a curve for each part of the impacted area (or a generalized curve for the entire area) showing how magnitude varies through time, or to construct a map showing how magnitude varies across the area at a given time. Picture, for example, a series of topographic maps showing the mine pit at different times. In practice what is drawn is, at best, a crude approximation in only two dimensions, of a complex multidimensional relationship.

In the following example magnitude is measured by population size.

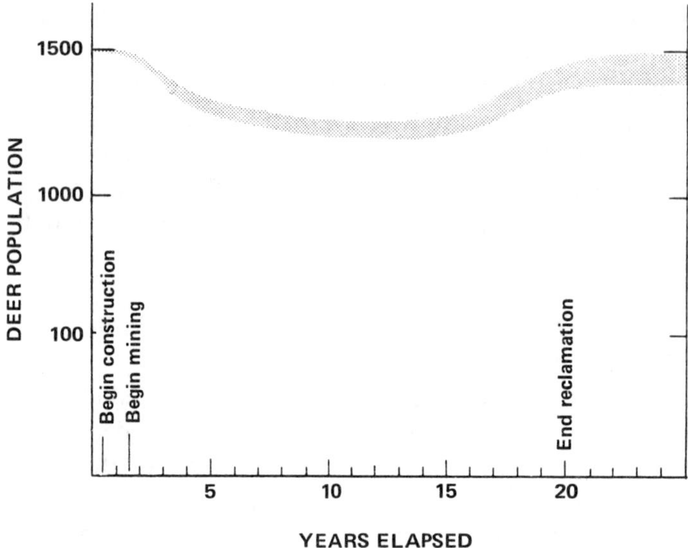

FIGURE 15-2. Changes in deer population due to mining. Width of shaded area suggests uncertainty of the forecast.

During construction and the first few years of mining, the Antelope Creek deer herd would probably decline rather rapidly from its present level of about 1,500 deer. For the remaining 20 years of mining, it would probably stabilize at around 1,300. After reclamation, it would probably increase gradually to about 1,400 to 1,450 — a few dozen below the pre-mining level (Figure 15-2).

For many disciplines, one probably can draw such a curve, and with hard thinking, find a way to pin numbers on its axes. Rarely will they be hard numbers, and the "curve" may well be a fuzzy band, but it should be possible to state whether the magnitude of the impact would change gradually or abruptly and to estimate whether the magnitude would be low, moderate, or high.

Significance. It is important to recognize that assigning significance is subjective; ultimately, it depends on value judgments. It is also important to recognize that the need to assign significance is dictated by NEPA, which requires discussion of "actions signficantly affecting the . . . . environment." Perhaps it is most important of all to recognize that for a federal EIS the significance assigned to an impact should be from a national perspective and that a locally significant impact may not be nationally significant.

Suppose that the topography in our proposed mine were progressively (time dependence) lowered an average of about 40 ft (magnitude) across the entire mined area (spatial dependence). How much would that matter? Or, more specifically, how much would a change of that magnitude interfere with anticipated post-mining land use? If the anticipated use is grazing, and if grazing would be practically the same after mining as before, the change would be of relatively little significance. If, however, mining would eliminate rock niches that were the unique habitat of the endangered hoary bedthrasher, the change would be of relatively great significance. (Society values endangered species enough that it has passed laws saying you mustn't harm them.)

Realistically, one can probably distinguish at least two levels of significance — say, "relatively little" and "relatively great." It might be possible to even distinguish between little, moderate, and great, but it is probably unreasonable to try to go further.

Generally, significance can be ascertained by asking, for each resource in the impacted area, "To what extent would the impacts degrade (or enhance) the things that society values?" Impacts of the proposed coal mine on soils would probably be insignificant, since within a decade or so after mining, the soils would probably support essentially the same uses they now support; and post-mining land uses would probably be essentially the same as at present. Impacts on air quality, however, could be highly significant, since concentrations of particulates at the east (downwind) boundary of the leasehold could violate both federal standards and Montana guidelines by a factor of two or three, several times per year, for the entire 40 years of mining.

There are numerous yardsticks for measuring significance. Perhaps the most obvious is the standards set by law (as for air or water quality). But commonly, other yardsticks are devised to suit specific situations.

ALTERNATIVES

Alternatives are all the means of achieving a goal. Decision making consists of formulating and selecting one of them, using certain criteria to make the selection. To do its job, the EIS must formulate the alternatives so the decision maker can choose among them.

How do you choose among alternatives? Suppose it's raining and your objective is to go about business as usual. Table 15-1 lays out some of the things you'd probably want to consider in deciding what to wear. The table shows the impacts of four alternatives on four issues. We call them "disciplines" since in an EIS, these issues would be topical disciplines — air quality, soils, vegetation, wildlife, socioeconomics, etc. As with most EIS's, adverse (undesirable) impacts or environmental costs are expressed as +'s; the much rarer beneficial impacts as –'s, negligible impacts as 0's. A +++ impact is severe, a ++ impact moderate, and a + impact slight (disregard, for the moment, whether it's magnitude or significance we're considering).

REVIEW OF THE DRAFT ENVIRONMENTAL STATEMENT

The required review period for each DES offers the public the opportunity to influence a federal decision, by ensuring that the decision is based on all relevant factors and that those factors are presented accurately and objectively. Ideally, review of the DES would be purely objective. But being human, we have biases. People biased toward development tend to look for

TABLE 15-1
DECIDING WHAT TO WEAR IN THE RAIN

	Alternative			
Discipline	Umbrella	Raincoat	Bathing suit	Street clothes
Wet	++	+	+++	+++
Hot	0	++	––	–
Encumbered	++	++	0	0
Ugly	0	+	+++	0

+ , adverse impact; 0 , negligible impact; – , beneficial impact

what they regard as overstatment of impacts. People biased against development tend to search for errors that result in understatement of impacts. This isn't as cynical as it sounds, for there is a statistical balance between views; a roughly equal number of reviewers with biases for and against development tends to produce an objective FES.

The excerpt below is from a recent EIS and includes bad text along with the good, to convince you that when text is hard to understand the fault isn't necessarily all yours — unless, of course, you wrote it. It also exemplifies the sort of deficiencies you might look for in reviewing a DES. The questions that follow are based on the excerpt.

Description of the Proposed Federal Action

> The mining and reclamation plan proposed by Consol must comply with applicable Federal, State, and local laws, regulations, and operating procedures to be approved by the Department of the Interior. In particular, compliance with 30 CFR 211 (Coal Mining Operating Regulations), 42 F.R. 3642 (1977) (Cooperative agreement with the State of Wyoming for enforcement of surface coal mine reclamation standards), and 30 CFR 700 et seq. (Surface Mining Reclamation and Enforcement Provisions) is required. The Pronghorn plan was written before the provisions were law and was filed prior to passage of the Surface Mining Control and Reclamation Act (SMCRA). However, compliance with all laws, regulations, and operating procedures, whether included in the present Pronghorn plan or not, will be required before the plan can be approved. Therefore, required modifications of the plan to comply with the law are part of the proposed Federal action. Accordingly, the following description is divided into a summary of Consol's proposal and a discussion of the constraints upon this proposal.

1. In the excerpt, who is proposing what? Answer: Consol is proposing to open a strip mine for coal.
2. What is the federal action? Answer: To consider the proposed mine plan, to decide whether to approve it, and if approved, with what stipulations.
3. Is the federal action proposed? Answer: No.
4. We suspect you found these questions hard to answer on the basis of the information provided (which is complete and only slightly modified from an actual (EIS). Why did you have trouble? Answer: The text doesn't provide the information needed to answer the questions.

EXERCISE XV

ENVIRONMENTAL IMPACT STATEMENTS

Name ______________________________

1. When the XX mine opens, the number of workers will increase very rapidly during the 2-year construction period to a maximum of 225, and then decrease somewhat more gradually to about 185 — a level that will remain roughly constant for the duration of the 25-year mine life. Please prepare a graph showing population versus time to illustrate this statement (recognizing that the information you've been given is incomplete).

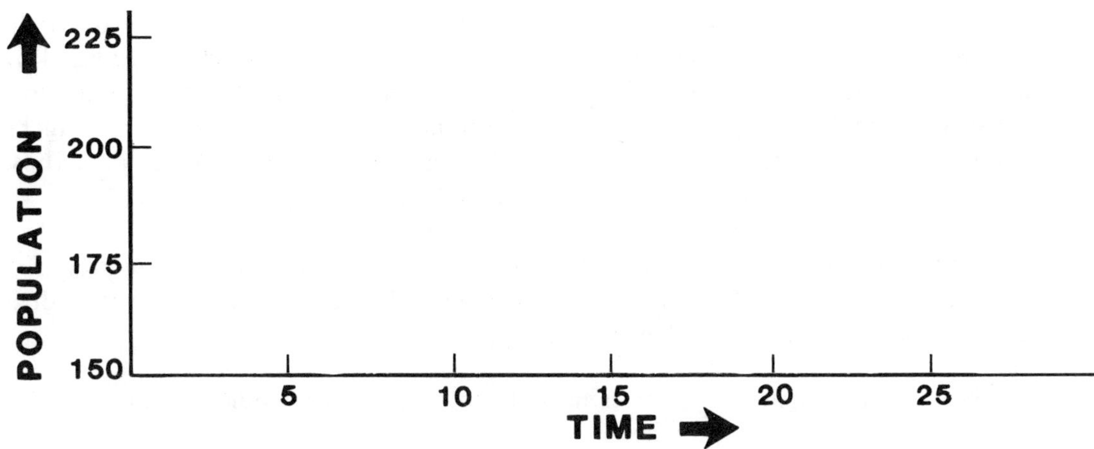

2. The cluster of x's in map t_1 below shows an area that has been impacted by the removal of vegetation. Please complete the series of maps, showing a larger impacted area farther to the northeast for time t_2 and a small impacted area still farther to the northeast for time t_3. Assume that at time t_2 most of the impact at t_1 has been eliminated by revegetation, and similarly, that at t_3 most of the impact at t_2 has been eliminated.

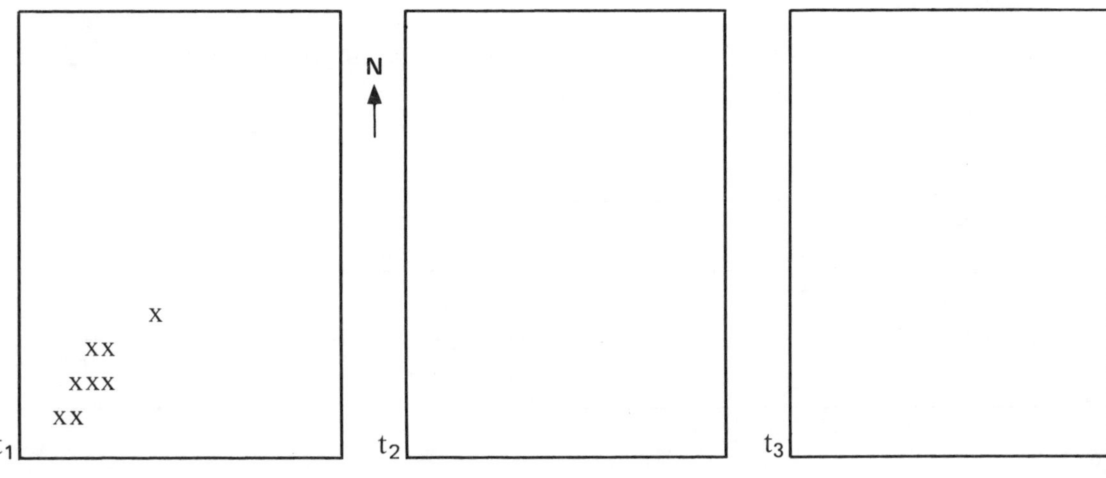

3. On the basis of the excerpt below, do you agree that the impact to antelope is major? Why or why not? ______________________________

____

____

____

____

The antelope density per square mile ranges from 18 during the summer to 31 during the winter. These figures suggest that about 35 to 60 antelope would be displaced by the mine. The number of antelope affected would be between 0.1 and 0.2% of the total of 32,000 antelope in Campbell County. The primary food for antelope, big sagebrush, would be eliminated for the short term, reducing their source of forage and, thus, the number of antelope that the reclaimed land could support.

4. Given that a 5-year storm is one that you'd statistically expect to occur once every 5 years, and given that a typical large western coal mine is worked for 20 to 40 years, does the statement below use a reasonable yardstick? Please discuss.

> We consider impacts on erosion insignificant, in that they would probably be less than those due to a 5-year storm.

____

____

____

____

5. How would you measure the significance of an impact on ground water? ______________________________

____

on wildlife? ______________________________

____

on recreation? ______________________________

____

6. Please consider the following excerpt.

> Mining would have a major impact on mineral resources. During the 22-year mine life, 100 million tons of coal would be removed at a rate of 2 to 5 million tons per year. The mine would extract 2% of the strippable coal reserves in the county.

(a) Remembering that every proposal has some purpose, is it meaningful to say that the removal of the coal would have a "major" impact on mineral resources? Please discuss. ______________________________

____

____

____

____

(b) Insofar as the question is ambiguous, what further information would you need to put things into context? ______________________________

____

____

____

____

(c) One possible way to measure the significance of impacts of a proposed coal mine is to compare them with the impacts of a hypothetically perfect mine. How significant would removal of the coal be in this context? Please discuss.

____

____

____

____

7. Refer to Table 15-1.
(a) If you attached equal weight to each "discipline" (wet, hot, etc.), which alternative would you select?

____

(b) Which alternative would you select if your overwhelming concern was:

to stay dry? ______________________________

to be neither hot nor cool? ______________________________

(c) Which alternative would you select if staying dry was twice as important to you as

any other discipline? ______________________________

If it was four times as important? ______________________________

Note that comparisons across the rows of the matrix are valid and readily made. The "wet" row tells you that you get wetter with an umbrella than with a raincoat, and in a bathing suit or street clothes you get absolutely soaked. The problem comes in making comparisons within a single column: with an umbrella, are you hotter than you are wet? That's basically a nonsense question. Sense can be made of it only by suitably weighting each discipline. And you've just seen how treacherous that can be.

(d) We evaded the question of whether the entries in the decision-making matrix represented the magnitude or the signficance of the impacts. What are the advantages of each?

____

____

____

8. On the basis of the following text, please discuss the effect of a proposed mine on insects: when, where, and how severely would they be affected?

____

____

____

____

Trace elements tend to accumulate in insects, especially at the higher trophic (feeding) levels. Many of these elements are toxic at low concentrations, depending on insect size, species, mobility, and feeding or behavioral patterns. Honeybees may be the most susceptible of animals to fluoride, arsenic, lead, copper, zinc, phosphorous, mercury, cobalt, and cyanide. It is believed that honeybees and other pollinators may accumulate these elements from pollen. This makes honeybees very susceptible targets for trace element toxicity even when these elements are found at very low levels in vegetation and soils.

9. In EIS's it is commonly necessary to hedge, that is, to qualify statements so they are no more positive than the available supporting information warrants. Please give a series of words or phrases that express the likelihood of an impact, from extreme unlikelihood to virtual certainty.

____

____

____

____

10. Please set up a matrix of your own for buying a new car. Select a preferred alternative for the discipline. (You'll probably want to assign +s to benefits of each car.)

	Alternative				
Discipline	Audi	Buick	Cadillac	Datsun	Edsel
Fuel Economy					
Noise					
Performance					
Reliability					
Safety					
Styling					

11. Young Enterprises is proposing to build a large chemical plant 2 miles northwest of campus. Without consulting your colleagues, please assign relative weights to the following disciplines. Make your own decision as to the maximum weight you wish to assign to each discipline — 1, 5, 10, 98, 837... After you've decided on weights, please rank the disciplines in order of the weights you've assigned (1 = greatest weight, 7 = smallest).

	Discipline						
	Air Quality	Geology	Soils	Vegetation	Wildlife	Socio-economics	Esthetics
Weight							
Ranking							

Now, compare your weights and rankings with those assigned by your various colleagues. To see how much variation there was in your colleagues' rankings, give the range in rankings for each discipline. To see how widely the assigned weights varied, give the range of weights for each discipline.

	Discipline						
	Air quality	Geology	Soils	Vegetation	Wildlife	Socio-economics	Esthetics
Range of Weights							
Range in Rankings							

As a potential scientist or decision maker, what does this tell you about assigning weights in the decision-making process? ______________________________

____

12. In selecting tracts to lease for coal mining, the decision maker commonly has to choose a certain number of tracts out of more possibilities. If he wants to select two tracts out of five possibilities, the number of alternatives (in this case, pairs of tracts) he must consider is 10. Suppose he wants to select three tracts out of ten possibilities. The number of alternative trios he must consider is 120. Just intuitively, how many alternatives could you intelligently compare? (Remember the problems you experienced in assigning relative weights to the various disciplines.) ____________________

13. Read the summary below from the draft environmental statement for the Antelope coal mine in Converse County, Wyoming, and answer the questions that follow. (See also the cover sheet for this DES in Figure 15-1).

DESCRIPTION OF THE PROPOSAL

NERCO proposes to open the Antelope surface coal mine in the eastern Powder River Basin, Converse County, Wyoming, 55 miles north of Douglas. The mine plan area, comprising of 7,642 acres of gently rolling topography, is used for ranching and wildlife. NERCO would extract 260.3 million short tons of low-sulfur subbituminous coal during 29 years, and in the process would disturb 5,860 acres. The mine would contribute as much as 12 million tons per year toward the Department of Energy's annual coal production goals . . . . NERCO presently has contracts to supply 5.6 million tons per year.

NEED FOR FEDERAL DECISION

NERCO has submitted a mining and reclamation plan (MRP) to comply with the State of Wyoming's regulations . . . Because NERCO has fulfilled the requirements of its leases, a decision on the proposed mine plan and permit application is required by law. . . . The Secretary (of the Interior) could approve the

MRP if its identified deficiencies are corrected (alternative A); disapprove the MRP (alternative B); or take "no action" (alternative C).

IMPACTS OF THE PREFERRED ALTERNATIVE (ALTERNATIVE A)

The impacts of the mine would be comparable to those of other surface coal mines in the area. The mine (about 9 mi^2) would be a small part of about 400 mi^2 of changing landscape due to energy development in the area. When viewed in terms of the eastern Powder River Basin (7,780 mi^2), the impacts on the mine site become insignificant. However, these impacts add to the effects of all other coal mining activities in the Powder River Basin. Thus, the significance of the impacts from the Antelope mine depends in large part on the perspective.

Following is a brief summary of the impacts of the proposed mine:

During mining, the geologic strata up to a maximum depth of 250 feet, soils, and vegetation on as much as 5,770 acres of the mine site would be progressively excavated. Changes in stratigraphy would be permanent. Soil structure would be radically changed but would reform over time. The vegetative productivity and the use of the land (presently, grazing and wildlife) should be restored to premining levels within 40 years after revegetation begins. Although vegetation would support a land use similar to that before mining, postmining vegetation would be somewhat different from the existing vegetation. Mining would remove part of one antelope range. Although there would be a loss of wildlife during mining, it is expected that wildlife populations would return to normal following reclamation. The lowered ground-water levels due to mining of the coal aquifers and to pumping water from a deeper aquifer would still be noticeable 30 years after mining ceases, and may take up to 70 years to reach normal conditions. Existing wells that only partially penetrate the overburden aquifer might have to be deepened to fully penetrate the aquifer: however, the cost impact of deepening these wells would be minor compared with the increased cost of pumping the water. The quality of the ground water in the replaced overburden would not be as good as present water quality (680 to 2,690 milligrams per liter (mg/1) total dissolved solids (TDS), but would still be acceptable for anticipated postmining uses (3,000 to 5,000) mg/1 TDS). NERCO is required to replace any developed water supply affected by their mining.

Significant cultural resources, including 48 sites nominated to the National Register, have been found in the mine area. A mitigation plan for documentation and recovery of these resources has been found acceptable by the State, OSM, and the Forest Service.

Despite some degradation of air quality in the vicinity of the mine, the quality of the air would still meet Federal and State standards. Traffic to and from the mine would cause and encounter short delays, increase the noise level, and add to maintenance requirements of the roads. If blasting is done in accordance with regulations, only short-term minimal impacts would result. When the mine is at full capacity, 23 100-car unit trains would convey coal from the mine each week. This would increase the noise, and the possiblity of accidents and delays at crossings along the railroad south and east of the mine.

By 1990, the mine would contribute about 1,069 and 1,345 persons, respectively, to the population increases in the city of Douglas and in Converse County resulting from energy development. Such population increases would place a heavy demand on these two governments (particularly that of Douglas) to provide services, such as police, fire, water supply, waste disposal, health, recreation. Because the mine would increase the property valuation, Converse County generally would be able to meet the demands placed on it. Douglas, however, would not have an immediate source of additional income, and already is experiencing difficulty in financing improvements. The applicant's socioeconomic mitigation proposal, submitted to the Wyoming Industrial Siting Commission and OSM, would mitigate some of these impacts.

Mining and associated facilities would constitute a visual intrusion to the generally rural landscape during the construction and active mining phases. This instrusion would be, for the most part, local in that the minesite is located on the south slope of a hilly area away from major viewing areas. After reclamation, most of this impact would be removed.

IMPACTS OF OTHER ALTERNATIVES

Alternative B: Disapproval. Under this alternative, the adverse environmental impacts associated with the approval alternative would not occur. Disapproval would be necessary should the applicant not correct the deficiencies in the MRP identified by the Wyoming Department of Environmental Quality.

Alternative C: No Action. The impacts due to the approval alternative would not occur if "no action" were taken. Any mining or other resource uses needed to meet energy demands would have to occur elsewhere.

(a) Given that the land has been leased for coal mining, that the Antelope Coal Company has submitted a proposed mine plan, and that the decision maker must now select one of the three alternatives considered in the EIS, do you feel that the three alternatives are sufficient? Why or why not?

____

____

____

(b) What other alternatives might have been considered?

____

____

____

(c) Do you agree that there would be no impacts if the proposed plan were not approved? ______________________________

Appendix A
Sources of Geologic Information

MAPS AND CHARTS

Maps and charts prepared by the U.S. Geological Survey are sold over the counter by all Public Inquiries Offices of the survey. Maps of areas east of the Mississippi River, including Minnesota, Puerto Rico, and the Virgin Islands, may be obtained from:

Branch of Distribution
U.S. Geological Survey
1200 S. Eads Street
Arlington, VA 22202

Maps of areas west of the Mississippi River, including Alaska, Hawaii, Louisiana, Guam, and American Samoa, may be obtained from:

Branch of Distribution
U.S. Geological Survey
Box 25286, Federal Center
Denver, CO 80225

Residents of Alaska should order directly from:

Branch of Distribution
U.S. Geological Survey
Federal Building, Box 12
101 12th Ave.
Fairbanks, AK 99701

Public Inquiries Offices sell maps of local areas and book reports of local and general interest over the counter. They are also depositories of selected open-file reports (not published) pertaining to the area served. They can be most helpful in researching everything from camping areas to old ghost towns.

ALASKA
108 Skyline Bldg.
508 Second Ave.
Anchorage, AK 99501

CALIFORNIA
7638 Federal Bldg.
300 N. Los Angeles St.
Los Angeles, CA 90012

CALIFORNIA
504 Custom House 555 Battery St.
San Francisco, CA 94111

Room 122, Bldg. 3
345 Middlefield Road
Menlo Park, CA 94025

COLORADO
169 Federal Bldg.
1961 Stout St.
Denver, CO 80294

DISTRICT OF COLUMBIA
1028 General Services Bldg.
19th and F Sts. NW
Washington, DC 20244

TEXAS
1C45 Federal Bldg.
1100 Commerce St.
Dallas, TX 75242

UTAH
8105 Federal Bldg.
125 S. State St.
Salt Lake City, UT 84138

VIRGINIA
302 National Center,
Room 1C402
Reston, VA 22092

WASHINGTON
678 U.S. Court House
West 920 Riverside Ave.
Spokane, WA 99201

Appendix B
Conversion Factors

Length		Volume	
1 micron (μ)	.001 millimeter	1 cubic inch (in.^{3})	16.4 cubic centimeters
1 millimeter (mm)	1000 microns .1 centimeter .001 meter	1 cubic foot (ft^3)	1728 cubic inches 28.32 liters 7.48 gallons
1 centimeter (cm)	10 millimeters .394 inch 10,000 microns	1 cubic centimeter (cc)	.061 cubic inch
1 meter (m)	100 centimeters 39.4 inches 3.28 feet	1 liter (l)	1000 cubic centimeters 61 cubic inches 1.06 quarts or .264 gallons
1 kilometer (km)	1000 meters 3280 feet .62 statue miles	1 cubic meter (m^3)	10^6 cubic centimeters 264.2 gallons 1000 liters
1 inch (in.)	25.4 millimeters 2.54 centimeters	1 cubic kilometer (km^3)	10^9 cubic meters 10^{15} cubic centimeters .24 cubic miles
1 foot (ft)	30.5 centimeters .305 meters		
1 mile (mi)	5280 feet 1760 yards 1609 meters 1.609 kilometers		

Mass		Area	
1 kilogram (kg)	2.2 pounds 1000 grams	1 square inch (in.^{2})	6.45 square centimeters
		1 square foot (ft^2)	144 square inches
1 metric ton (mt)	2205 pounds 1000 kilograms 1.1 tons	1 square centimeter (cm^2)	.155 square inch 100 square millimeters
		1 square meter (m^2)	10^4 square centimeters 10.8 square feet
1 pound (lb)	16 ounces 454 grams .45 kilogram	1 square kilometer (km^2)	247.1 acres .386 square mile 10^6 square meters
1 ton (tn)	2000 pounds 907.2 kilogram .91 metric ton		

Appendix C
Geologic Time Scale*

Eras	Periods	Epochs	Age estimates, million years before present
Cenozoic	Quaternary	Holocene	0.011
		Pleistocene	1.5–2
	Tertiary	Pliocene	11
		Miocene	25
		Oligocene	40
		Eocene	60
		Paleocene	70
Mesozoic	Cretaceous		136
	Jurassic		190–195
	Triassic		225
Paleozoic	Permian		280
	Pennsylvanian		320
	Mississippian		345
	Devonian		395
	Silurian		430–440
	Ordovician		500
	Cambrian		570
	Precambrian		4,500±

*after U.S. Geological Survey

Appendix D
Geologic Map Symbols

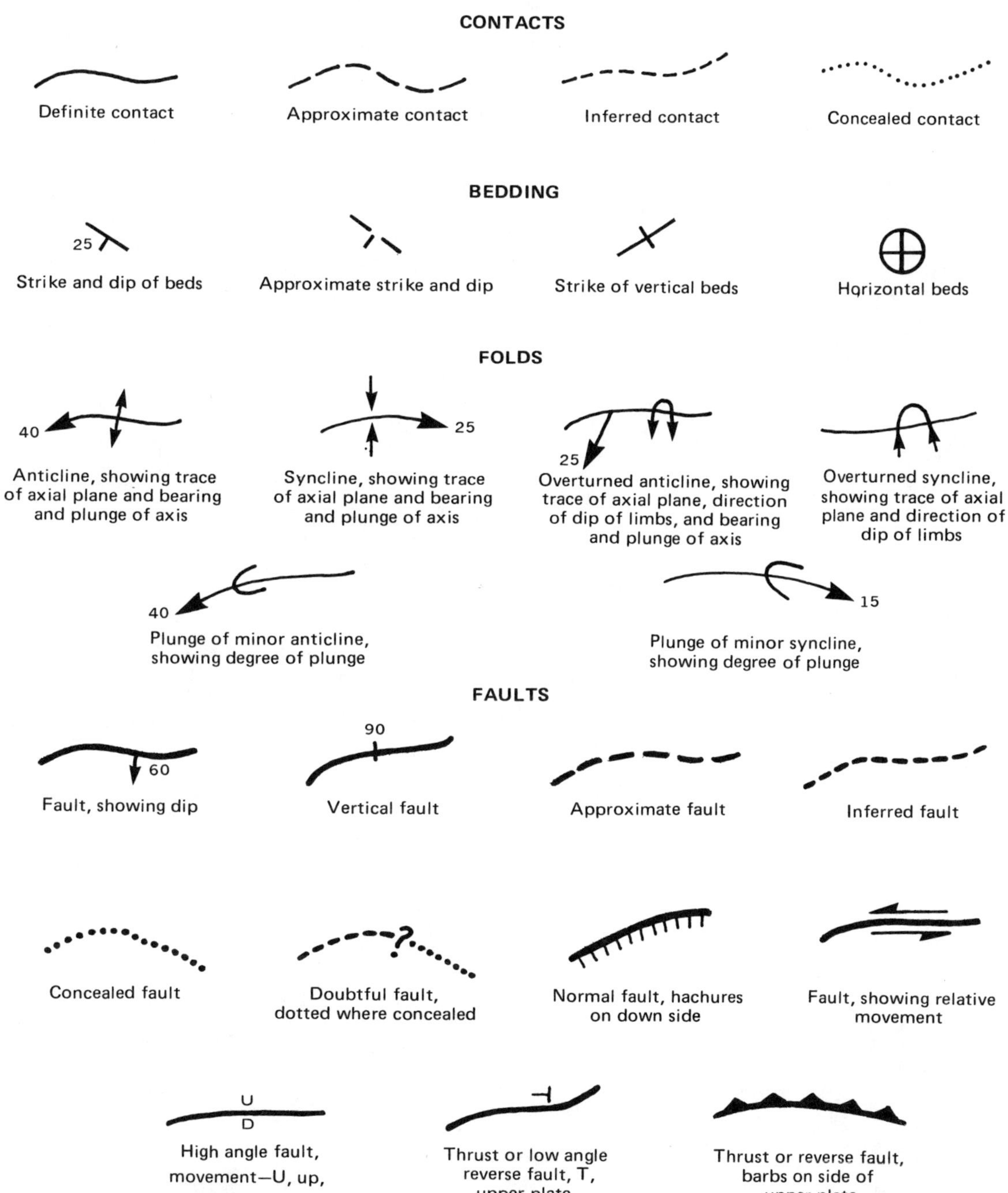